Also by Roland Ennos

The Science of Spin: How Rotational Forces Affect Everything from Your Body to Jet Engines to the Weather

The Age of Wood: Our Most Useful Material and the Construction of Civilization

Trees: A Complete Guide to Their Biology and Structure

THE
POWERFUL
PRIMATE

How Controlling Energy
Enabled Us to Build Civilization

Roland Ennos

Scribner

New York   Amsterdam/Antwerp   London
Toronto   Sydney/Melbourne   New Delhi

Scribner
An Imprint of Simon & Schuster, LLC
1230 Avenue of the Americas
New York, NY 10020

First Scribner hardcover edition February 2026

SCRIBNER and design are registered trademarks of Simon & Schuster, LLC

Simon & Schuster strongly believes in freedom of expression and stands against censorship in all its forms. For more information, visit BooksBelong.com.

For information about special discounts for bulk purchases, please contact Simon & Schuster Special Sales at 1-866-506-1949 or business@simonandschuster.com.

The Simon & Schuster Speakers Bureau can bring authors to your live event. For more information or to book an event, contact the Simon & Schuster Speakers Bureau at 1-866-248-3049 or visit our website at www.simonspeakers.com.

Interior design by Kathryn Kenney-Peterson

Manufactured in the United States of America

10 9 8 7 6 5 4 3 2 1

Library of Congress Control Number: 2025946884

ISBN 978-1-6680-6279-1
ISBN 978-1-6680-6282-1 (ebook)

Let's stay in touch! Scan here to get book recommendations, exclusive offers, and more delivered to your inbox.

To Peter Lucas, a good man in academia

Contents

CONTENTS

Prologue

THE POWERFUL PRIMATE

One scene above all others from the BBC's 2011 series *Human Planet* demonstrates how humans have come to dominate the natural world. In the film, three members of the Dorobo people of Kenya, a client tribe of the better-known Maasai, intimidate and steal meat from a pride of fifteen lions gorging on the corpse of a wildebeest. After stalking up unseen to within fifty yards, the three men suddenly stand up straight and march shoulder to shoulder directly toward the corpse. Startled, the lions scatter and retreat, like cats being chased from a bird they had just caught, and crouch at a distance, growling in the long grass. When the three men reach the wildebeest, two stand sentry, while with his knife the third swiftly slices through the skin, muscles, and ligaments of the animal's hip and pulls away a whole haunch of meat from the corpse. Within seconds, before the lions can decide what to do, the men are striding away again with their trophy; they have stolen enough meat for a feast without any of the danger or bother of a hunt. And later that evening they light a fire and cook the meat, gorging on the soft flesh and bone marrow.

To urban people from the developed north, used to a sheltered life in "civilization," the brazenness of this daylight robbery seems almost suicidal. How could a few weak primates intimidate a far greater number of the world's most feared predator? The men should be no match for lions, who could easily outpace them and could, being far stronger, wrestle them to the ground and dispatch them with a single bite to the neck. The physical superiority of lions over humans has been demonstrated throughout human history—the Romans even exploited it for entertainment. In one of their cruelest forms of punishment, *damnatio ad bestias* (condemnation to beasts), they threw convicted criminals—most notoriously Christians—to the lions. In front of crowds of thousands of spectators in amphitheaters such as the Colosseum in Rome, the condemned people met an inevitable and painful death.

The truth is, though, that ever since the advent of farming, we have tended to exaggerate the threat posed to us by large predators such as big cats, wolves, and bears. Over thousands of years of history, our propaganda has repeatedly sought to portray ourselves as the plucky underdogs, surviving against huge physical odds by marshaling our intelligence and native cunning against the brute forces of nature. Folktales are filled with people fighting to survive in forests, steppes, and deserts, and evading and outwitting the wild beasts that dwell within them. So when we venture into the wilderness, we do so with trepidation and we take a whole host of precautions. We keep predators away during the day by traveling in groups, maintaining constant vigilance, and employing firearms, while in the darkness of the night, we cower together around our campfires.

But in fact wild animals fear human beings much more than we do them, and indeed they fear us more than they do even the fiercest of predators. For instance, recent research by Liana Zanette of Western University, London, Ontario, and colleagues has shown that tape recordings of human voices startle a whole host of large African mammals, from warthogs to giraffes. The voices create almost double the effect on animals as the roars of lions. Indeed, lions themselves retreat when they hear human voices, and the reasons why lions fear humans are not hard to find. The success of the Dorobo raid on the lions' kill was possible because humans have consistently defeated lions over the long history of conflict between our two species on the East African plains. The pastoralist Maasai people, for instance, who also live in the region, have successfully been vying for supremacy with lions for hundreds of years. Armed with just simple wooden fighting sticks and short wooden spears, they can protect their herds of cattle from predation. And they don't just defend themselves and their cattle passively; they frequently take the fight to the lions and kill them. Until recently a successful solo lion hunt was a rite of passage for Maasai boys. Lions in the region had every reason to fear people, who could dispatch them with their handheld weapons and bows and arrows long before the advent of European big game hunters equipped with rifles.

The dominance of humans over lions was also exploited by the Romans themselves for another of their "entertainments," the *venatio* or wild-beast hunt. In these unedifying spectacles, the lion was just one of a host of large creatures, from wolves to giraffes, that were released into an amphitheater to be hunted down by specialist gladiators known as *bestiarii*. Being trained in combat, and armed with swords, spears, and bows and arrows, the gladiators were able to turn the tables on the wild beasts; the result was another predictable slaughter, in this case of the animals.

The demand for such entertainments and the Romans' efficiency at procuring and transporting wild animals to Rome was so high that they effectively wiped out the wildlife of North Africa where they sourced them. That the Romans were able to maintain a constant supply of wild beasts to their capital also shows how they dominated the whole of the North African landscape. Their engineering expertise had enabled them to build an empire controlling the entire Mediterranean region, feeding the huge city of Rome with cereals grown over vast swathes of farmland in Egypt, Tunisia, and Libya. Their huge grain ships and extensive system of roads were not surpassed in size or sophistication until well into the eighteenth century.

But perhaps the most impressive aspect of the Dorobo video was that it could be made at all and be distributed to millions of people worldwide. To shoot the video, the film crew had to fly by jet plane from Europe to Africa, travel around the plains of the Maasai Mara in four-wheel-drive Land Cruisers, and use sophisticated cameras made on the other side of the world to record the hunt and the Dorobos' subsequent feast. And once the crew had flown home again, they had to edit the film with sophisticated computer software and distribute it by broadcasting it as electromagnetic waves through the ether, or streaming it across the wires of the internet. Finally, the viewers had to pick it up on equally complex TV receivers, home computers, or mobile phones. All of these stages required a global infrastructure, huge technical expertise, and vast amounts of energy. We could see what happens in the wilds halfway around the world only because of our modern industrialized society.

And our dominance over the natural world is now more or less universal. Today humans are spread over every continent except Antarctica, and we have modified the world's surface beyond recognition. We have cleared over 40 percent of the habitable land for our agriculture, and we manage over half of the 31 percent of land that is still covered by forests. Our domestic animals outnumber wild beasts by a ratio of fifteen to one, and 26 percent of species of large mammals are threatened with extinction. And even before the advent of agriculture, humans helped eliminate large animals all over the globe: mammoths and woolly rhinoceroses from Europe and Asia; horses and mastodons from North America; giant ground sloths and armadillos from South America; and giant wombats and giant kangaroos from Australia.

The reasons for this domination are simple. Rather than being the feeble wimps we portray ourselves as, who have to make up for our inferior physiques with superior intelligence and cunning, we have long been the bullies of the natural world. In some ways we are actually physically the most powerful animals on the planet. Using the simplest of tools—sticks and stones—we can hit harder, throw farther, and cut deeper and cleaner than any other animal, giving us unprecedented abilities as scavengers and predators of animals, and as harvesters of plants. We can kill animals even at a distance, efficiently skin and dismember them. We can cut down and uproot plants; crack open their nuts and grind down their seeds; strip their leaves and separate and spin their fibers. We can fell trees and carve wood. We can build fires and cook food. We can clear forests, till the soil, and mine the rocks. In other words, even using just our muscles we can engineer almost all aspects of our environment. It is our ability to marshal our physical power to produce energy and to concentrate it using our tools that first enabled us to remake the world for our own convenience. And in the last few thousand years we have even learned how to use our tools to co-opt power from other sources: from wood and charcoal; from draft animals; from water, wind, and the sun; and most recently from fossil fuels and atomic nuclei. This book charts the history of humanity by tracing the path by which we progressively increased the power and energy we could generate; improved our ability to concentrate and apply it; and magnified the distances over which we could transmit it.

It may seem surprising to attempt to understand the progress of such complex organisms as human beings simply by examining our physical relationship with the world around us. Most other human histories have tended to concentrate on the finer aspects of the mind and the rise of culture. However, as we shall see, this physical perspective makes sense of human history in a way that no other viewpoint achieves. It allows us to understand why it was a primate that gained world dominance rather than any other type of animal. And it allows us to integrate the roles of the many other factors that anthropologists and historians have implicated in our rise to dominance: human bipedalism; the evolution of our opposable thumb; our development of stone tools; the growth in our brain size; our increased sociality and lengthened adolescence; the rise of agriculture; the inventions of metallurgy and machinery; and the harnessing of fossil fuels, steam, hydraulics, and electricity. To me the engineering approach also seems to be the most logical one. After all, the only way we can interact with the world around us is by converting energy from one form into another, and to do it at a reasonable rate: in other words, to do anything we need to generate, transmit, and apply power. And though we often think of ourselves as purely intellectual beings, quite separate from the natural world, we are still animals and, like all other organisms, are subject to the laws of physics and natural selection.

One reason I can tell this story now is that we have a better understanding than ever before about how the human body works. Over the last half century, the science of comparative biomechanics, in which I have been involved for over forty years, has been remarkably successful in explaining the design and function of animals and plants—how animals stand up, walk, run, climb, jump, eat, for instance, and how this has affected the evolution of their bodies and minds. For instance my friend Peter Lucas has used the science of fracture mechanics to explain why mammals have such varied designs of teeth. In recent years, anthropologists and primatologists have also started to investigate the evolution of the human body, and the development of our tools, in much the same way. Medical researchers and sports scientists have also been investigating the mechanics of human movement for just as long, though for reasons of their own, and they have studied humans in isolation from research on other animals. In their definition *biomechanics* refers exclusively to humans. Fortunately, though, anthropologists take a broader view and are realizing that it is time to integrate our knowledge of human motion into evolutionary studies, so we can better understand how human beings emerged from our ape ancestors.

Meanwhile, over the last fifty years primatologists have also made important discoveries about how our primate relatives construct their sophisticated sleeping nests, and how they make and use mechanical tools such as probes, levers, spears, and digging sticks. These findings invalidate many of the old assumptions about how humans evolved. In particular they destroy the concept of "man the toolmaker." We now need a new narrative to explain instead how humans capitalized on and improved our tools—how we became better engineers—and so shed light on how humans evolved into scavengers and hunters.

Anthropologists and historians alike are also starting to reassess the causes and effects of the emergence of farming. They used to assume that settling down to grow cereals and keep animals was a key advance in human progress. Recently, however, historians such as James C. Scott and Yuval Noah Harari have characterized farming as a misstep that condemned our ancestors to lives of drudgery and privation. It is time to examine these competing narratives in a quantitative way to work out why after the last ice age people in different parts of the globe adopted such different ways of life as hunter-gatherers, herders, horticulturalists, and cereal farmers. Despite what we are invariably told, cereal farming is not the best way of making a living at all, but it did stimulate farmers to become better engineers: to domesticate and harness draft animals, to develop metallurgy and devise machinery. In so doing it sparked a technological revolution that led to the invention of wheeled vehicles and plank ships and resulted in the advent of large states and empires, changing patterns of supremacy across Eurasia, and leading to the ultimate triumph of the Old World over the New.

Nowadays, we also have the benefits of fifty years of research into the archeology of the industrial world. It is time to rescue industrial archeology from its neglected silo among engineers and men in boilersuits to demonstrate the crucial importance of engineering in the formation of our modern world. It is time to investigate why it happened when and where it did, and how engineering advances led us to increase our power and use more and more energy. These insights allow us to understand how industrialization shaped modern history, and how it led to urbanization, the transformation of the countryside, and to globalization.

And since all engineering involves destruction, it is time to investigate the increasingly dramatic effects of human power on the horrors of war and on the health of our planet. Understanding our past and present may help us plan a sustainable future.

I believe that my interests and lifetime's experiences have given me the ideal background to tackle the task of building this new synthesis. I was fortunate enough to be able to combine my childhood fascinations with animals, airplanes, and bridges to forge a university career researching and teaching in the field of comparative biomechanics—the engineering of animals and plants. I spent three happy decades investigating such subjects as how insects fly, how plants anchor themselves in the ground, and how grasses use glass to defend themselves against being eaten. I carried out research on the engineering of the human body, investigating how our fingernails are designed; why we have fingerprints; and how we initiate walking. I devised research projects that related to the evolution of humans: how apes can move safely in the forest canopy and how they build their sleeping nests; and why early humans fire-hardened their arrows and how they designed their stone axes. I gained practical experience in agriculture by investigating how to prevent cereal crops from being blown over and how best to separate the fibers of flax from the stem of the plant. And together with my partner, I maintained an allotment, a plot of land for cultivation, for several years. Outside work, I have spent many happy days visiting archeological sites, and open-air and industrial museums, marveling at the beauty and ingenuity of the buildings, tools, and machinery. Better than any academic education, these living museums help us picture the lives of ordinary people, how they lived and how they worked to transform the world into what it is like today.

Throughout those thirty years I have gradually tried to assimilate and integrate all this information into books about the natural world and the place of human beings within it. Following textbooks on the environment and plant life, I wrote a popular book on trees. And more recently, especially since I took early retirement, I have had the time to tackle human history itself. In *The Age of Wood* I pointed out the crucial role that this one material has had on the human story. And in *The Science of Spin* I showed for the first time how rotation has a pivotal role in the way people walk, run, throw, and swing their tools, as well as in the machinery they developed to power the modern world.

In this book, my aim has been to combine all of this information to build up a comprehensive history of how people have exploited their physical prowess and engineering skills to build ourselves a safer, more comfortable world. I am sure it will not be the last word on the subject; I realize more and more as I get older just how little I know. But I hope that it might stimulate others—primatologists, anthropologists, archeologists, and historians in particular—to take the engineering aspect of our lives more seriously. And I also hope it will supply people with enough information to put us human beings into our proper historical context so that we can understand ourselves better and help us escape from the current predicaments in which we find ourselves of climate change and unhappiness. Above all I hope that I can persuade you that our physicality and engineering ability are integral aspects of what makes us human and have shaped the world in which we live; I want to show that they are not only worthy of study, but of awe and respect.

Part One

DEVELOPING OUR POWER

Chapter 1

THE DAWN OF POWER

If the scenes of the Dorobo intimidating lions on the Kenyan plains or of Roman *bestiarii* massacring wild beasts in the arena show the extent of modern humans' dominance over the natural world, another, more peaceful, scene may perhaps point to its birth. For fifty years ago in the forests of the Taï National Park in the Côte d'Ivoire, Hedwige and Christophe Boesch of the University of Zurich first observed chimpanzees using lumps of wood and stones as hammers to crack open nuts. The ape places the nut onto an "anvil," a large stone or the knee of a tree root, before raising a wooden or stone "hammer" and bringing it down on top of the nut, splitting its shell and releasing its contents. The ape then gathers up the pieces of shattered kernel and eats them, before breaking the next nut in its hoard.

The whole process looks almost human; the apes resemble nothing so much as bank clerks stamping stacks of documents. And fifty years ago this human quality shocked the anthropological establishment, which had built up the evolutionary just-so story that humans had "broken away" from the apes thanks to our unique ability to make and use stone tools. This story had started to break down in 1964, when the pioneering primatologist Jane Goodall discovered that the chimpanzees of the Gombe National Park, Tanzania used grass stems as fishing sticks to extract termites from their nests and mosses as sponges that

3

they dipped into pools of water to help them drink. Anthropologists started to realize that they had to abandon the simple concept of "man the toolmaker." They either had to redefine humanity or think of new ways in which we are different from the rest of the animal world. At first it seemed possible to take the stance that humans were different because we alone use tools made of stone. The finding that chimpanzees use stone hammers was the final blow even to this definition.

To work out the real ways in which we are different from the rest of the animal world, we first have to understand where we have come from: to examine what we have inherited from our closest relatives, the primates, and in particular from the great apes. If we do that, we can see that the key aspect of the lives of primates that differentiates them from other mammals is that they live high up in the forest canopy and have, over the past 50 million years, become adapted to an arboreal existence. This has made them very different from terrestrial mammals, and as we shall see, it also preadapted both them and us to make and use tools.

The selection pressures acting on the early primates, living high up in trees, would have been totally different from those on land mammals. Free from the fear of being eaten by land-based carnivores, they no longer needed to be able to run swiftly. They were able to slow down their metabolism and live longer, slower-paced lives, while their bodies adapted to help them move about more efficiently and safely through the trees. The first and most obvious differences between primates and terrestrial mammals is in the design of their limbs. In land mammals the front and hind legs are usually similar in form and function; they both swing backward and forward. To enable the animals to run quickly, to trot and to gallop, they tend to be as slender as possible, and they have the muscles concentrated at the base to minimize the energy required to move them; their fore and hind feet are both stretched to elongate their strides and end in hooves or claws.

Unlike land mammals, the primates differentiated their limbs, each pair having contrasting structure and function. They evolved sturdy hind legs that could support them from below and freely mobile arms that could reach out to grasp branches or hold on to the trunks of trees and pull themselves upward. Their limb muscles stretch well down their length, particularly those in their arms, which gives primates the strength they need to climb. Above all, primates developed broad hands and feet with finely articulated fingers and toes, and an inner surface that is covered in fleshy pads, which are patterned with prints and which are backed by nails. These adaptations improved their ability to grip on to narrow branches and twigs and made their fingers uniquely sensitive. This enabled primates not only to move rapidly and safely through the tree canopy, but also suited them to reach out and pick the two sources of food that are plentifully available in the forest canopy and which grow at the tips of branches—fresh new leaves and ripe fruit. Fruit is an especially good food for primates such as apes because natural selection has designed it to be attractive and nutritious. Primates eat the fruit and digest it, but leave the tree seeds inside intact so that they are dispersed in their feces. The flesh of fruit is full of sugars and other nutrients, and as it ripens, it softens, enabling primates to bite into it with their wedge-shaped incisors and chew it into a digestible pulp with their sharply cusped molars. Apes such as chimpanzees can even test whether a fruit is ripe by palpating it between their fingers, like the archetypal French housewife inspecting produce at the greengrocer's.

Primates also enhanced their vision to enable them to thrive in the treetops. Tree fruits often advertise their palatability by changing color from green to yellow or red to show that they are ripe. Primates therefore evolved better color vision than terrestrial mammals, and they moved their eyes from the side of the head to the front, giving them binocular vision so that they could judge distances better and so help them make their way more quickly and safely through the canopy.

But it was not just the primates' bodies and senses that underwent a thorough redesign. Just as important to their success—especially that of the largest of these creatures, the apes, which emerged in the past 30 million years—were their large brains. Most primatologists emphasize the social benefits that large brains confer on apes, since they may help them cooperate with each other in increasingly elaborate ways, while enabling the most intelligent individuals to manipulate their peers. Yet there is no obvious reason why this behavior should evolve in arboreal mammals such as apes rather than in terrestrial ones such as cats. A more immediate advantage of a large brain for arboreal animals is that it helps them deal with the physical world around them. It allows them to memorize not only where fruit trees grow in their territories but at what time of year each tree fruits. This means that fruit-eating macaques and apes can compute traveling routes to the nearest fruiting trees and arrive at them just as the fruit is ripening, and before potential competitors such as hornbills and squirrels.

Even more important to the apes as they increased in size was that their larger brain also enables them to judge the strength and stiffness of branches before they risk putting their weight on them. This skill is essential to allow such heavy animals to reach the fruit hanging from the tips of slender twigs, and to travel safely through the canopy, especially to cross the perilous gaps between trees. A mistake could lead a heavy ape to plummet to its death. The apes' need to compare their weight with the strength of branches and to consider the consequences of their actions on their environment might also have led them to evolve a degree of self-awareness; apes are among the few animals capable of recognizing themselves in a mirror.

Together, the physical and mental capacities of apes—the strength, mobility, and dexterity of their arms, hands, and fingers, and their awareness of the mechanical world around them—have even enabled them to develop their own technology. All the current great ape species (and so presumably our own ancestors) evolved the ability to build themselves complex sleeping nests. Every evening an ape decides on a suitable tree to spend the night in and chooses a thick branch on which to build its nest. My former PhD student Adam van Casteren showed that it goes about its task with speed and precision and demonstrates surprising engineering know-how. Leaning out, it grabs hold of more slender branches and half breaks them, bending them inward and weaving them together into a cup-shaped cradle in which the animal can sleep securely. Finally, it breaks off still thinner branches and sprays of leaves to make a soft mattress and to cover itself with a blanket of foliage. Nests enable apes to get a good night's sleep, away from predators, and prevent them from tumbling to the forest floor. From the sophistication of these structures, and the speed at which apes can build them—an orangutan can construct one in just a few minutes—it is clear that apes have become familiar with the unique properties of wooden branches and other plant structures and are adept at exploiting them. They clearly merit the accolade of being first-rate architects and engineers.

For many apes, that is about as far as they go in their engineering careers; after all, their hands and teeth are well adapted for their diet of fruit, so they have no need to invent feeding tools. But their mechanical skills are merely dormant and can be called on whenever needed. Orangutans, for instance, which live in the lofty canopies of rainforests in Southeast Asia, rarely use tools in the wild. However, in zoos they are notorious for dismantling any device placed in their cage and effecting the most ingenious escapes from their enclosures. And when they find themselves in more challenging habitats where there are fewer fruits to feed on, wild orangutans do develop a toolmaking culture. In the swamp forests of Suaq, Sumatra, for example, they use two sorts of tools made from simple sticks: one set to extract honey from bees' nests in the hollow trunks of trees; the other set to prize open the hard shells of *cemengang* fruit. The apes even modify the design of their shell-opening tools through the season, choosing wider, stronger sticks later on as the fruit's cases stiffen up.

Life is not so easy for the African apes. The formation of mountains along the Great Rift Valley has, in the past 20 million years, started to cut off the rain-bearing clouds of the monsoon from the rainforests of central Africa. As a result, tree cover over the continent has become more fragmented, and the African apes, especially our closest relatives, the chimpanzees and bonobos, have had to descend more often to the forest floor and to develop a wider range of tools to obtain enough food. The commonest tools are the fishing sticks that many apes use to probe into holes in termite mounds to extract the insects. Recently Adam van Casteren was part of a team who showed that to achieve this task chimpanzees choose especially flexible plant stems, which can go around bends more easily so the chimps can push them deeper into the holes. These tools mimic the tongues of pangolins and woodpeckers. In contrast, most of the tools that apes use are much stouter wooden sticks that function as levers or bayonets and mimic the way other animals use their hard, rigid claws or tusks. These tools concentrate and magnify the forces the apes' muscles produce to enable them to break into their food. The chimpanzees of Gabon, for instance, carry around with them a whole wooden tool kit to break open bees' nests and raid them of honey. This includes stout sticks to pierce the nest; lever-like enlargers to widen the hole; slender collectors with frayed ends to dip into the honey; and "swabbers," elongated strips of bark, to scoop it out, like the tongue of a bear.

Life is most difficult for chimpanzees in the savannas that have developed at the driest eastern and western edges of their range, where there are few trees, and where the dominant vegetation is grasses. In such habitats it is particularly hard for an ape to make a living. Savannas are home to large, fast-moving carnivorous mammals, such as lions, leopards, hyenas, and wild dogs, so they are far more dangerous places than the forest canopy. The food is also more difficult to find, eat, and digest. Some savanna trees bear fruit, but in the majority of species, the seeds are protected from herbivores by rigid nutshells.

And the grasses and herbs that carpet the savanna floor don't produce fruit at all, only small, hard seeds, while they hide their stores of energy underground in bulbs and swollen roots.

Field research over the last thirty years has found that savanna chimpanzees have responded to this challenge by developing a whole new suite of tools. The savanna chimps in Tanzania, East Africa, make digging sticks twelve to twenty-four inches (30–60 centimeters) long, with which they probe into the soil to excavate and lever out plant tubers. The chimps living in the savannas of Senegal, West Africa, make tools that are even more familiar to us; females construct spears. They break off two-to-four-foot-long (60–120 centimeter) branches, strip them of their leaves, and use their teeth to sharpen the narrow end into a point. They then probe the spears into hollow tree trunks to flush out and, if possible, impale small primates such as bush babies hiding within the holes.

But one major savanna food source—nuts—would be inaccessible to apes that relied on strength alone and tried to use static forces to break into them. The mouths of apes do not open wide enough to enable them to bite into many nuts, and their jaw muscles and teeth are not strong enough to crush them even if they could. Their hands and fingers are also too weak and their finger pads too soft to make any impression on their surfaces; and their nails are not rigid enough to prize them open. Of course, most of us have found around Christmastime that it certainly *is* possible to break into nuts by concentrating force. We use nutcrackers: tools that are specially designed to use the principle of the lever to produce forces large enough to crush nutshells. But the forces involved are huge. My experience is that hazelnuts and walnuts are not too difficult to crack, but even they require us to mobilize most of our gripping force, around fifty pounds (200 newtons), to squeeze the arms of the nutcracker together. Since most nutcrackers operate at a mechanical advantage of over 5:1, this means that these nuts require a force of over 250 pounds (1 kilonewton) to break them.

Almonds and Brazil nuts are even harder to crack; as a child I learned to limit my losses and leave them to the grown-ups. They must take up to five hundred pounds of force (2 kilonewtons) to break.

Apes could theoretically make and use nutcrackers, since their grip strength is typically twice that of ours, but nutcrackers are extremely hard to construct. The jaws have to be made of especially hardened steel so they do not deform when used; and the two metal arms of the device need a strong hinge joint. Apes would never be able to construct them. In any case nutcrackers have disadvantages as well. Nutshells are fairly smooth, so the nuts are all too apt to fly out of the jaws of the nutcracker. And because the forces needed to break the shells are so great, once fracture begins, it is hard to reduce the pressure fast enough; the jaws are all too apt to accelerate destructively inward, snapping shut and causing catastrophic fracture not only of the shell, but also of the kernel within—crumbs of nut fly uncontrollably in all directions.

As we saw at the start of this chapter, the Taï chimpanzees (along with several other populations in West Africa) have devised an alternative technique to crush nuts and have invented a totally different sort of tool—a hammer. Though it may seem simple, this primitive tool and the novel way in which it is used mark a significant development in chimpanzee technology. The great majority of primate tools act as levers or sharp points, magnifying and concentrating the forces produced by the arm muscles. Primates use the tools in the same way that other animals use their teeth or tusks. Lions kill their prey by concentrating the forces produced by their jaw muscles at their pointed canine teeth. Warthogs and elephants dig by using their tusks to localize the force they exert on the soil. Even herbivores use force to break off edible pieces of plants. Cows eat grass, for example, by grasping the leaves between their teeth and tongue and pulling them away from the base of the plant.

But muscles can be used for other purposes than generating large forces. They can also shorten at the same time as they are producing force, generating mechanical energy. To crush nuts the Taï chimps rapidly contract their triceps muscles, straightening their arms and accelerating their forearms downward, building up kinetic energy in their hands and in the hammer stone. What happens to this energy when the stone hits the nut depends on the mechanical properties of the two objects. As the two objects collide, they deform, and the dynamic force between them rises rapidly to a peak that is much higher than a chimpanzee could apply simply by pressing down on the stone. As the force of the collision rises, the two objects both deform, but not by the same amount. The more compliant object—the nut in this case—deforms far more than the more rigid stone, so almost all of the kinetic energy within the stone is transferred into the nutshell. The subsequent fate of the nut depends on the fracture properties of the material of which it is made. If it were made of a tough material such as wood, the nut would absorb the energy without breaking; its internal structure would merely be slightly damaged. But nutshells are brittle. They are made of strong, thick-walled cells, but unlike wood cells, these are short; and lines of fracture can easily run between the cells. Cracks can also race through the shell along the special lines of weakness that many nuts have evolved to allow them to split open when they germinate. (This is why walnuts often split along the seam down their center when you crack them.) So when a Taï chimpanzee hits a nut with their stone hammer, the shell breaks, releasing its contents, while the hammer stone remains intact.

Breaking nuts with hammers may seem to be but a minor step toward dominating the planet. After all it is just one of the many ways in which apes use tools, and chimpanzees themselves are hardly numerous, so it has not enabled them to achieve world domination. Moreover, only a few populations of chimpanzees have ever learned how to use hammers. One reason for this must be the chimps' lack of mechanical insight. Though you might think it would be simple for chimpanzees to learn how to crack nuts, the skill has only been mastered by a few chimpanzee populations. It takes many months for chimpanzees in these groups to perfect the technique, and juvenile apes have to observe their elders and put in around two years of practice before they achieve success. Chimpanzees who have grown up in a group that does not crack nuts never learn the skill, even when primatologists provide them with the nuts and the necessary tools. The idea that they could generate huge dynamic forces using muscle power to build up kinetic energy in a tool seems to be hard for them to grasp.

This should not be surprising. Despite centuries of science, few people, including many primatologists, and anthropologists, appreciate the difference between force and energy, and the difference between static and dynamic forces. But fortunately, most people do manage to build up an intuitive feel for the benefits of percussion tools. We realize that we can do more damage by hitting things than merely pressing on them, though most people would be hard-pressed to say why. Consequently, as we shall see in the next few chapters, our ancestors made widespread use of such simple percussive tools. For instance, though nuts are usually seen as a minor source of food, hunter-gatherer societies around the world have used hammer stones to crack open and exploit nuts on a surprisingly large scale. Many of the Neolithic settlements around the Atlantic coast of Scotland, for example, are surrounded by huge mounds of hazelnut shells, which must have been harvested in their thousands. In California, the Cahuilla tribe developed a whole "balanoculture," which was dependent on the harvesting, shelling, leaching, and grinding of acorns. In coastal areas around the world hunter-gatherers also used hammer stones to break open the shells of edible mollusks such as mussels, whelks, and limpets. And our ancestors subsequently improved the hammering technique in a whole range of ways, which eventually led to the emergence of humans as efficient scavengers and ultimately as top predators.

Chapter 2

PUTTING OUR BACKS INTO IT

If you watch any of the numerous detective dramas that clutter the television schedules—and they are so common, it is possible to watch two or more every night—you should be a veritable expert in murder by blunt instrument. In the first scene of many of these shows, the victim is found with the back of their skull stoved in, by a stone, paperweight, or statuette depending on the location and social standing of the victim. Two hours later, after a whole shoal of red herrings have been cleared away, the murderer is finally revealed and confesses that in a confrontation they just "lost it," picked up the nearest heavy object, and dealt the victim a single, fatal blow to the back of the head. The reality is little different; it actually is quite easy to kill people in this way; for instance eleven murders in England and Wales in 2022 were inflicted by blunt objects and thirty-five by hitting or kicking without a weapon. In the USA, even though more lethal weapons such as guns are freely available, blunt instruments still accounted for 317 victims in 2023.

The lethal effectiveness, when armed with a blunt object, of even relatively weak humans—the killer is just as likely to be a slight woman as a burly man—is at first sight astonishing. After all, as we saw in the last chapter, our cousins the chimpanzees can struggle to crack open nuts with stone hammers. This is despite that apes have far stronger arms than we do; they need them to haul their bodies up one-handed when they are climbing in the forest canopy. The arm muscles of apes make up a far larger proportion of their total musculature, around 31 percent, than in modern humans, at 18 percent. And their muscles are also composed of a greater proportion of strong fast-twitch fibers. So when primatologists performed those long-term behavioral experiments back in the 1950s and '60s, bringing up baby chimps in their own homes to see if they could teach them to speak like human beings, the animals quickly became too strong to handle. They could not be managed by their tutors, and most were carted off to zoos where they lived out the rest of their lives free from the horrors of a human education. The resolution of the apparent paradox, that a weak human can deal a much more powerful blow than a much-stronger ape, is down to the ways in which we have learned to use our bodies and contract our muscles—it's what we do with them that counts!

As we have seen, the actions of nut-cracking chimpanzees are extremely simple. An ape swings its forearm down using the most intuitively obvious method: it contracts its triceps muscles to extend its elbow joint and straighten its arm. This technique certainly works, but it limits the amount of power and energy that can be produced, which is related to the characteristics of its power source: muscle.

Muscle is an extraordinary tissue, capable both of producing force and of generating power. Structurally, it consists of two alternating sets of long fibrous protein molecules, actin and myosin, which stick out on each side of a series of plates, like bristles from two-sided shoe brushes. In the muscle cell, the sets of fibers overlap, just like when you stick two shoe brushes together, and the fibers can be joined side by side by large numbers of releasable crossbridges that link the myosin fibers to the actin fibers. To produce a large contractile force, the crossbridges are all joined and bend back, like a person's finger pulling back on the trigger of a gun. This tends to shorten the muscle and can produce forces of up to forty pounds per square inch (0.3 MPa). When you are producing a constant force, such as holding up a heavy weight, this is what your muscles are doing. You are performing no actual work, but the action still requires energy, because each crossbridge detaches every now and then and has to be reattached and flexed again, using up fuel as it does so.

To generate power and produce mechanical energy, the muscle has to shorten at the same time as it is producing its contractile force. To do this the crossbridges have to release at a faster rate and swing forward before they reattach, allowing the fibers to slide past each other, before they reattach farther along and bend back again, rather like an inchworm crawling along the ground. The faster the muscle shortens, the more crossbridges have to be detached at any one time, so the less force the muscle can generate; consequently force declines more or less linearly with contraction speed. Eventually, at the maximum shortening speed of a muscle (typically two to four lengths per second), it can produce no force at all.

To produce the maximum amount of energy over a single contraction, a muscle needs to shorten slowly. The force it can generate in this way is more or less the same as it produces when it is held still, in what muscle physiologists call an isometric contraction. And since a muscle can shorten by around 25 percent of its initial length, when it contracts slowly, it can produce around seventy joules per kilogram over the whole contraction. However, contracting muscle in this way produces a minimal power output, since it takes so long to generate the mechanical energy. To produce a maximum power output, a muscle needs to contract much faster, at around 35 percent of its maximum speed, at which point it can produce just 35 percent of the maximum isometric force. Under these conditions a muscle can generate a power output of eighty to one hundred watts of power for each kilogram of muscle, but each contraction can produce much less energy than if it had contracted slowly, just twenty joules per kilogram.

As we have seen, when they are cracking nuts, Taï chimps straighten their arms by contracting their triceps muscles. Unfortunately these muscles are relatively small because the apes rarely need them to produce power when they are moving in the trees, in contrast to the biceps, which they use to bend their arms and so haul themselves upward. Consequently their triceps muscles are 25 percent smaller than the biceps and weigh only around sixteen ounces (450 g), compared with 24 ounces (680 g) for the biceps. Contracting the triceps muscle can therefore only give a peak power output of thirty to forty watts, and the triceps can generate just nine joules in each contraction. This would be enough to crack open a walnut, which only takes one to two joules to break, but the nuts the Taï chimps eat, produced by *Panda oleosa* and *Coula edulis* trees, have far stronger shells, which take between twelve and one hundred joules to crush, depending on the type of hammer used. The blow the chimps produce should never be able to crack open these nuts.

The chimps get over this by adding to the energy produced by their muscle action another form of energy: gravitational potential energy. They don't use small stones as hammers as you might expect, but logs or large rocks, which weigh on average twenty-two pounds (ten kilograms). They lift them slowly using their powerful biceps and shoulder muscles, until they are between twenty and forty-seven inches (50 and 120 centimeters) above the ground, storing 50 to 120 joules of potential energy, before the apes let them drop. The chimps therefore crack nuts not so much by using muscle power as gravity; by holding on to the rocks as they fall they merely supplement gravity and control their descent. This controlled-dropping technique is also used by several species of small capuchin monkeys and macaques, which can crack open a range of nuts and shellfish. Being far smaller than chimpanzees, between two and eleven pounds (one and five kilograms) in weight, these monkeys have even more difficulty providing the energy they need. They achieve their aim by lifting as large a stone as they can in both hands and using their leg, back, and arm muscles to raise it high above their heads. This gives the stone as much potential energy as possible, before the monkey finally brings the stone down on the nut, like the defender of a medieval castle flinging a rock down on a besieger.

Using such massive stones works, but it has major drawbacks. A twenty-two-pound (ten-kilogram) weight is hard for a chimpanzee to lift, ungainly to handle, and virtually impossible to transport. The apes have to have a permanent nut-cracking site where they bring their nuts. And a chimpanzee could never use a twenty-two-pound (ten-kilogram) tool as a weapon with which to hit other chimps, or as a projectile to throw at predators or prey; it is far too heavy. As a tool such a rock is single-use and an evolutionary dead end.

So how is it that humans, who are much weaker than chimpanzees, manage to use smaller stones to stove in other people's skulls or throw them as projectile weapons? The answer is that we have devised methods to accelerate our tools that simultaneously use a great many more muscles. This has enabled us to produce far higher bursts of muscle power than apes, and to transfer much greater amounts of kinetic energy to our tools. We generate this power by exploiting what I call sling action.

Most people, including sadly most biomechanics and sports scientists, seem to believe that the only way to swing our limbs is to activate the muscles around our joints, to directly rotate them. This is exactly what the Taï chimps do. However, there are other ways to swing our limbs. The most obvious way to do this is to use the force of gravity, swinging our arms and legs to and fro past vertical, like pendulums, as athletes often do during their warm-ups. This action can also be strengthened if we accelerate the proximal joint upward, as we do at each footfall when we walk or run. After our stance leg hits the ground, it accelerates our hips and body upward, producing an enhanced pendulum action that accelerates the swing leg rapidly forward to allow it to be ready to take the next step. This enhanced pendulum mechanism is especially strong when we run because the forces on the stance leg can be more than three times our weight. It can be even stronger when we kick soccer balls; we plant our stance leg firmly on the ground and push up with it, which accelerates our hip up and so swings our kicking leg rapidly forward through the ball. So when we walk, run, or kick, most of the energy to swing our legs comes from contractions of muscles in the other leg and in our hips. We swing our legs without needing to contract the muscles within them, which is why we can walk and run with such a graceful loose-limbed gait.

And there is also a third way to accelerate a limb segment: to move its proximal joint in a circle, so that it is accelerating inward. This produces a centrifugal force on the outer limb segment, creating a sling action that can accelerate it forward even faster than the enhanced pendulum action. This sling action, powered by the rotation of our thighs and upper arms, accelerates our lower legs and forearms when we walk. It straightens our arms and legs automatically toward the end of each step and produces the huge forces that accelerate our feet rapidly forward when we kick soccer balls.

We also use this same sling technique when we perform a hammer action such as banging our fists on a table. We don't just straighten our elbow, as a Taï chimp does; we also use our shoulder muscles to rotate our upper arm downward. This causes our elbows to move in a circular path, producing a sling action on our forearm, which aids the action of our triceps muscle to straighten our elbow and accelerate our hand. Using this technique, we activate not only the ten ounces (300 grams) of muscle in our triceps, but also the four pounds (2 kilograms) of internal rotator muscles in our shoulder. Just by using our upper arms, we can produce five times more power and energy than the chimpanzee.

We also make use of major differences in our body shape and in the way we move to further multiply the power we can produce. We are bipedal and have much broader shoulders than chimpanzees and a longer, more flexible back. Consequently, we can use the muscles in our legs, hips, and back, including the largest muscle in our body, the gluteus maximus, the muscle that makes our bottom stick out, to rotate our shoulders. This adds a second stage to the sling action. It accelerates our upper arm faster than our shoulder muscles could do unaided. And the faster our upper arm swings, the more strongly the circular movement of our elbow powers our forearm, and so the faster we can swing our hands.

Another consequence of using this double sling action is that when we perform a hammering motion, we swing our hand much farther than a chimp does when it performs the same action. We move it from being well behind our body at the start to well in front of our body on impact, a distance of around two yards (two meters). This gives our muscles enough time to contract, even when we are using smaller, lighter stones as hammers, so we can accelerate our hands to much higher speeds by the end of the strike. We can therefore use much smaller, more portable stones that weigh in the region of one to two pounds (0.5 to 1 kilograms) as tools. And we can swing them much faster.

The double sling action is so effective at accelerating small stones that we can even propel them fast enough to use them as projectiles. We can throw stones to deter predators or to inflict damage on our enemies, even when they are still some distance from us. We can transmit the energy we produce well away from our bodies.

I don't know of any scientific studies that have been carried out to measure the amount of power and energy we can produce when we are hammering with a simple stone, but it is certainly much more than the Taï chimps can achieve with their simple action. And we can make a good approximation of it by examining how effective we are at performing its related action: throwing projectiles. These movements are fortunately far better studied because of people's modern-day obsession with sports. Sports scientists have shown, for instance, that a baseball pitcher can throw a 5-ounce (145-gram) baseball at speeds of up to 100 mph (45 meters per second), transmitting around 115 joules of kinetic energy to the ball. It is not possible to throw heavier objects quite so fast, but an NFL quarterback can throw a 14.5-ounce (415-gram) American football at speeds of up to 60 mph (27 meters per second), giving it 120 joules of energy, while a European handball player can throw a 16-ounce (450-gram) handball at 50 mph (22 meters per second), giving it ninety joules of kinetic energy.

Of course, sportsmen are exceptional physical specimens and receive intensive training and coaching, so an ordinary person could probably only generate half that energy, around fifty to sixty joules. Even so, this is easily enough to break open a human skull. Experiments (presumably not on living subjects) have shown that this requires as little as fifteen joules. And generating sixty joules of energy and transferring it to a one-pound (0.45-kilogram) stone would be enough to propel it at a speed of 45 mph (20 meters per second), fast enough for it to fly up to eighty feet (24 meters). In contrast to human capabilities, when chimpanzees have been trained to throw overhand, they have been unable to propel stones at speeds greater than 15 mph (7 meters per second). Having given the stones a mere nine joules of kinetic energy, they are unable to throw them more than fourteen feet (4 meters)!

Humans can even produce powerful blows when we are not holding any tool at all in our hands—we can hit people with our closed fists. Boxers deliver two main types of punches: jabs and hooks. In the jab, boxers accelerate their hands much like a nut-cracking chimpanzee, by contracting their triceps to straighten their arm. Since a boxer stands sideways on and jabs with the hand closest to his opponent, the fist does not have far to travel, which allows a jab to connect within as little as 0.4 seconds. However, the maximum hand speed is low, typically only 15 mph (7 meters per second), giving our fist, which weighs around fourteen ounces (400 grams) (or twenty-eight ounces [800 grams] with boxing gloves), just eight to fifteen joules of kinetic energy. In contrast, with a hook the boxer uses their rear hand, so the fist has farther to travel. It typically takes 0.65 seconds to land, so is more easily evaded. However, in a hook boxers rotate the whole body to swing the shoulder forward in a curve, producing a more powerful action that helps accelerate the fist up to 30 mph (13 meters per second), giving it thirty to fifty joules of kinetic energy. This is enough to knock out the opponent if delivered in the right place, though in modern boxing the blow is cushioned by the boxing glove, which prevents fractures of the skull or hand. The uppercut can be even more lethal, since it harnesses the powerful trunk muscles to help accelerate the fist upward.

We can punch so hard that even untrained people can use their boxing abilities to escape from the jaws of huge predators by punching them in the face. In 2020 an Australian surfer, Mark Rapley, saved his wife from a ten-foot (three-meter) white shark at Port Macquarie, New South Wales, while in 2024 a Canadian cyclist fought off a mother grizzly bear in Anderson Flats Provincial Park, British Columbia. Most impressive of all was the exploit of eighty-four-year-old Dolores Boppel, from North Fort Myers, Florida, who drove off a seven-foot (two-meter) alligator that was attacking her shih tsu. Truly we are the most powerful hitters and throwers in the animal world.

Of course, we are not born with the ability to generate and control such large bursts of power, and to hit so hard and throw so far. It takes children many years of practice to develop the coordination and timing needed to throw balls any great distance and in the right direction; some people, especially those who lack interest in sporting activities, never attain the ability. But all of us can appreciate the importance that practice has to help us hone our skills simply by trying to throw a ball with our nonpreferred arm; we find that since we have never practiced throwing with this arm, our performance is usually abject; as we sometimes say, "we throw like a child."

In contrast to throwing, hitting seems to be a skill that almost all of us acquire, but perhaps that is down to our having had more motivation and practice in our childhoods, during our incessant disputes with our siblings and classmates. It should be apparent that even those of us who have grown up in cosseted urban environments have unique powers about which we are largely unaware. It is not surprising that wild animals instinctively flee from us, and that animal-film makers have to build elaborate hides or use expensive drones to obtain footage of creatures behaving normally.

So when and how did our ancestors acquire our hitting and throwing skills? Unfortunately, since biomechanics has not taken on board the concept of sling action in modern humans, anthropologists understandably know next to nothing about the physical prowess of early hominins. However, it is unlikely that our earliest ancestors would have been able to hit as hard as us or throw as far as us 6 million years ago when our lineage first separated from that of the chimpanzees and bonobos. But there are certainly good reasons to believe that they were better at hammering and throwing than modern chimpanzees. Chief among these reasons was that even the earliest hominins were bipedal. The humanlike design of their hips, with femur bent inward at the top, shows that by 5 million years ago creatures such as *Orrorin tugenensis* and *Ardipithecus ramidus* were capable of standing and walking upright. By 3.6 million years ago, fossil footprints and skeletal finds show that early australopithecines, such as *Australopithecus afarensis* had straighter, arched feet like our own and had perfected upright bipedal locomotion; they walked about just like us.

Being bipedal would have given them many advantages. Most obviously, being bipedal would have freed their hands to carry and use tools, but it would also have enabled them to wield them more effectively. They would have been able to rotate their hips, bodies, and shoulders like we do to accelerate their arms using a powerful double-sling action. However, despite having stronger arms than us, which they must have used to help them climb back up into the canopy for food and to stay safe at night, they would not have been able to generate such large bursts of power as we do. The shape of their hips shows that they had much smaller gluteus maximus muscles. This means that they would not have been able to run as fast as us. A recent simulation study led by Karl Bates from the University of Liverpool showed that their maximum speed would only be around 11 mph (5 meters per second) compared with 18 mph (8 meters per second) for modern humans. Another effect of their small gluteus maximus would be that they would not have been able to rotate their bodies and shoulders so strongly, and they would not have been able to exploit their shoulder rotation so effectively either because they had narrower shoulders; these still resembled those of the great apes, to enable them to raise their arms vertically and haul themselves up trees.

Nevertheless, the selection pressures on australopithecines to improve their hitting and throwing ability must have been strong, and between 4 million and 2 million years ago their shoulders and hips gradually became more human, suggesting that they steadily increased the power they could generate. And they no doubt used their growing abilities to do more useful things than just carry out impulsive murders. Being able to produce more power and generate larger amounts of kinetic energy than great apes, they could have used digging sticks in a more effective way than merely as levers; they could pummel them point-first into the ground, as modern hunter-gatherers such as the Hadza do today, before prizing soil out of the ground. This would have allowed them to dig faster and to unearth much longer roots from much deeper in the soil. In the same way, they could have produced far more powerful stabbing actions with their primitive spears. And they could have used both sticks and stones as millstones to crack nuts and break grass seeds into smaller pieces so that they were more easily digestible.

Though *Australopithecus afarensis* was vegetarian, later australo-pithecines could also have obtained a wholly new source of food by using stones to break open the long bones and skulls of herbivore carcasses and extract the nutritious marrow and brains. Like the human skull, it takes between thirty and a hundred joules of energy to break open the skull and long bones of ungulates, which should have been well within their capabilities. Species such as *Australopithecus africanus* would have been able to scavenge these inaccessible parts of mammal carcasses. Few predators are capable of breaking open these bones because their jaws do not gape wide enough to grasp a skull, and their jaws and teeth are simply not strong enough to crush leg bones. Lions typically leave a carcass after less than twenty-four hours because by then the flesh has started to rot, leaving behind a more or less intact skeleton. The australopithecines could then have approached the abandoned corpse and harvested its marrow and brain, which being protected inside the bone can stay fresh for up to two days. The newly found power of australopithecines would therefore have enabled them to exploit a totally new food resource, one containing fats that could provide enough energy to help them survive and nourish their rapidly expanding brains. It would have enabled them to move into a totally new ecological niche: to become effective scavengers of animal tissue as well as gatherers of plant tissue.

About 2.3 million years ago australopithecines were joined in Africa by another genus of hominin, our own genus, *Homo*. The first member, *Homo habilis*, was essentially similar to the australopithecines, with powerful arms that suggests that it still spent much of its time in the trees. But the fossil evidence shows that by the advent of its relative *Homo erectus* around 2 million years ago, hominins had attained the full power capabilities of modern humans. *Homo erectus* was the first truly terrestrial member of our group, and the shape of its pelvis suggests that it had an enlarged gluteus maximus, just like us. This would have enabled it to swing its legs rapidly back and forth and so break into a sprint. It also had broad shoulders and slender arms, which show that it would have been poor at climbing up trees. However, along with its enlarged gluteus maximus, this would have enabled it to swing its arms as quickly as we can to throw projectiles and produce powerful hammer blows. It would have been a more effective scavenger than *Homo habilis*. And it had also developed other characteristics that would have given it the ability to manipulate a much wider range of tools and fill an entirely new ecological niche.

Chapter 3

GETTING A GRIP

One of the downsides of using tools, as any DIY practitioner will know to their cost, is that they are apt to be as dangerous to the user as the things they are meant to work on. I have incurred a whole host of cuts, bruises, blisters, and muscle strains over the years thanks to my amateurish engineering activities, carried out both in the home and in the research laboratory. In view of these dangers, professional artisans need to be highly trained to handle the power they impart to their tools and to control how and where they apply the energy they generate. Because of the simplicity of their hammers, this would have been even more of a problem 5 million years ago for our early hominin ancestors, just as it is today for the Taï chimpanzees and their hammers.

The most obvious problem with using handheld hammer stones is that we could bash our fingers on the target, but there is another, less immediately apparent difficulty. During a hammer blow we build up kinetic energy not only in the stone tool, but also in our hand and arm. On impact all this energy needs to be dissipated, as well as the energy within the stone. Ideally most of it would be transferred, via the stone, to the target and be used to break it, but to achieve this our hands would have to be stiffer and harder than the target. And if this were the case and we had rigid hands, the impact would send massive shock waves through our wrists and arms, causing huge peak stresses that could easily fracture them. So it was fortunate that we inherited fingers and palms from our primate ancestors that are covered in soft pads of fat, pads that act as shock absorbers, and tough but flexible skin covered in prints that provides an excellent grip and resists blistering. But even with padded hands and fingers, the impact forces when a hammer stone hits its target can reach hundreds of newtons and can jerk the stone out of our hands, potentially causing a great deal of damage to everything around.

Unfortunately, the hands of modern apes such as the Taï chimps are not well adapted to provide the firm grasp necessary to withstand such powerful impacts and to keep hold of hammers. Their palms are slender, their fingers are long and curved, and they have short thumbs. The reason for this is that apes' hands are designed primarily for their arboreal lifestyle: climbing around the forest canopy. Apes typically hold their four fingers together, using them as a hook, which provides an admirable temporary attachment to a tree branch, while allowing them to easily let go as they swing onto the next branch. But because of the length disparity between the fingers and the thumb, they cannot hold firmly on to an object between the pads. The thumb is just too short, so chimpanzees either have to hold objects between the very tips of their thumb and fingers, or between the thumb pad and side of their index finger, just as we hold keys. Neither technique is secure, and it is all too easy for objects to fall from their grasp. The lack of an opposable thumb also means that apes are poor at grasping sticks: the shaft can slip between their fingers and the stick can rotate across their palm; if they try to swing it around, it will tend to fall from their grasp.

The wrists of African apes are also adapted to aid their locomotion rather than to hit or throw. Along large branches and on the ground they walk quadrupedally, resting the weight of their forequarters on their knuckles, so they have a reinforced wrist joint. Though they can flex it forward and inward, the joint resists being bent back, so their wrists stay straight when they put their weight on their knuckles. The hands of modern humans, in contrast, being free from the need to support us either in the treetops or on the ground, show many adaptations that help us hold on to and use tools. The most noticeable adaptation is that we have shorter fingers than apes and longer thumbs, so we can hold objects between the pads of our fingers and the pad of our thumb to generate a stronger grip. We also have other adaptations to improve the security of our grasp. Our fingers are straight, rather than being curved, and they end in broader, flatter pads or apical tufts, giving us a greater area of contact with our tools, enabling us to generate higher friction for a given grip force.

Together, these adaptations greatly improve the efficiency and versatility with which we can use our hands. First, they help us grasp spherical tools. We can hold a one-pound (0.45-kilogram) stone firmly in a secure "power grip," wrapping our fingers around one side and our thumb around the other, so that even a sharp jolt cannot wrest it from our palm. We can also wrap our hands into a fist, protecting our fingers during a punch and helping to more effectively transmit the dynamic forces up into our arms without damaging our metacarpal bones. We can even take firm hold of narrow objects such as twigs or pens, by grasping them between the pads of the thumb and two or more fingers, using a range of what anthropologists call precision grips.

Our wrists, too, are different from those of great apes. They are more flexible, and in particular we can extend them backward and outward as well as flex them forward and inward. This allows us to impart even more energy into a blow from a stone and to punch harder by adding a final third stage to our sling action. We can hold our wrist bent back at the start of the swing, before letting it swing forward under centrifugal force just before impact. We can also use this same wrist action to throw projectiles faster and farther. This is most noticeable in darts players, who swing their forearms in a controlled rotation from the elbow, like the Taï chimpanzees, to produce repeatable, accurate throws, but who generate much of the speed of their darts with a final flick of the wrist.

However, an even greater advantage of the way our hands have been modified through evolutionary history is that we can make far better use of long cylindrical tools, most notably sticks. We can hold them firmly in a modified power grip, with the fingers wrapped one way around the stick and the thumb wrapped the other way farther along the shaft. This secure grip allows us to resist greater forces along the axis of a stick, improving our ability to plunge it into the ground, spear an animal, or pound grain. But holding a stick in this way also gives us a much more useful ability: it allows us to swing it with extra power. We can produce torques with our wrist, enabling us to hold the stick steady at the start of the swing, so that it rotates along with our forearm. By the time we have built up speed, our wrist mobility then allows the stick to swing forward due to the centrifugal forces on it. The stick acts like a much longer, heavier hand, providing a much more powerful third stage in the human sling action. We can even strengthen this action by actively flexing our wrist toward the end of the motion. By wielding a stick in this way we can magnify the amount of energy we can give to it and deliver a much-harder blow than if we were just using a handheld stone. The stick can act as an effective club.

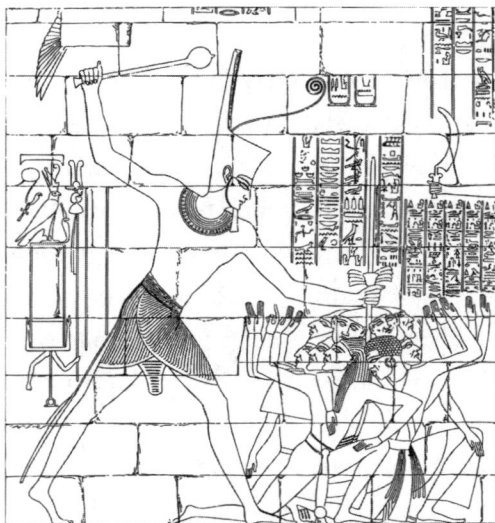

The pharaoh Seti I striking prisoners of war with his mace, from the temple at Karnak, Thebes. The Egyptian "side-on" style of art is particularly good at showing the way Seti is about to turn his shoulders, unleashing a three-stage sling action that accelerates his club. Note the power grip with which he is holding it.

Apart from magnifying the energy we can generate, using a stick or club rather than a handheld hammer has other advantages. First, it allows us to attack predators or adversaries while keeping them at a safer distance. Second, because the point of contact of a club is well away from our hand, we are far less likely to damage our fingers, and our palm receives far less of a shock on impact, so a club is a far safer tool to use. Indeed, if the impact is around two-thirds of the way along a club from its base, at what swordsmiths call its point of percussion and tennis racquet makers call the sweet spot, we experience no shock at all: all the energy of the blow goes into the target and none needs to be absorbed by our own body. The effectiveness of a club can be further improved by concentrating its weight in a heavy "head" at its far end. And a stick can also be used in another way: as a lethal projectile weapon—a throwing stick. Because the far end of a stick travels far faster than the hand, after it is released, the stick can travel much farther than a stone of the same weight; it somersaults through the air as it flies, delivering a far more devastating blow, and can hit a target much farther away.

Few experiments have been carried out on the mechanics of swinging clubs or throwing sticks. Fortunately, however, we have a good idea about how effective they can be because, once again, sports scientists have carried out extensive studies on sports stars swinging the modern equivalents of fighting clubs: tennis racquets and baseball bats. Films of professional tennis players have shown that in forearm shots and serves, they can generate racquet speeds on the order of 100 mph (45 meters per second). Since a racquet typically weighs around ten ounces (300 grams), this means it contains 240 joules of kinetic energy, over twice the amount we can give a simple projectile such as a ball or a stone. We can transfer more energy to the racquet because its head travels much farther than our hand, so our muscles have longer to accelerate it, and they can act at a more efficient, slower speed. Baseball batters, who hold on to their bats with both hands, and who can consequently mobilize even more muscles to swing them, can build up even more energy. They can swing their two-pound (900-gram) bats at up to 75 mph (34 meters per second), generating around 400 joules of energy. Whether a one-handed or two-handed club is used, a single blow generates far more energy than is needed to break the long bones or skull of a human or indeed of any large mammal, and so a blow from a club is likely to prove fatal.

Given their effectiveness, it is no surprise that clubs have been a favorite weapon of humans right up until living memory. Wooden fighting clubs, weighted at the far end and carved with elaborate decoration, were the weapon of choice of Polynesian warriors, the Maasai, among other groups in Africa, and several Native American tribes including the Mohawks and Iroquois. They were also popular in Japan until medieval times, and fighting sticks are even depicted in the Bayeux Tapestry, among more modern weapons such as battle-axes and swords. Throwing sticks were also popular throughout the ancient world. The ancient Egyptians hunted birds using sticks weighted at the far end, while the most effective throwing weapons of all must be the straight hunting boomerangs used by the Australian aborigines, which are flattened to reduce drag, and which can be lethal at ranges of up to 220 yards (200 meters).

Thanks to the benefits of being able to hold on to tools more firmly and swing them about more powerfully, there must have been strong selection pressure on our hominin ancestors to improve their grasp and free their wrist. We cannot be sure, however, how quickly the first hominins developed the grasping and tool-using capabilities of their hands because of the poor fossil record. Finds of complete hominin hands are extremely rare, and in any case it is difficult to relate structure directly to function, particularly when it comes to the complex architecture of hominin wrist bones. It is even difficult to determine the direction of evolutionary changes in the earliest of the hominins, since it is unclear whether the last common ancestor of humans, chimpanzees, and bonobos was bipedal, or whether it was a knuckle-walking quadruped. However, the evidence shows that by 4 million years ago, early hominins had already evolved shortened fingers and lengthened thumbs. The early australopithecines could probably grasp tools at least as firmly as we can, particularly as they had even broader pads on their fingers and thumbs than we do. On the other hand, they retained curved fingers, which would have facilitated tree climbing, and they seem to have had less flexible wrists, so they may not have been able to wield hammers and sticks as powerfully as we do. Nevertheless, Kubrick's vision in *2001: A Space Odyssey* of ape-men using bones as clubs for hunting and intergroup fighting may not have been so wide of the mark.

By 2 million years ago, the first totally terrestrial hominin, *Homo erectus*, had hands and wrists that looked and functioned essentially like ours, just like the rest of its body. *Homo erectus* would have been able to run, hit, club, and throw stones just as effectively as a modern human and use tools such as stone hammers, wooden clubs, and throwing sticks just as well. Together, these adaptations would have helped it to protect itself from terrestrial predators and may even have enabled it to drive them away from the corpses of the victims of predators. *Homo erectus* would have been an even more effective scavenger than earlier hominins. And as we shall see in the next two chapters, it may even have developed the ability to exploit more of the corpse, to eat a wider variety of animal tissues than just marrow and brains; and it would have become a predator itself. But before hominins could achieve this breakthrough, they had to improve not only their physique and biomechanics, but also their mental capabilities, so that they could make and use an entirely new set of tools.

Chapter 4

CUTTING IT

If you have watched any of those nature documentaries about wild-life on the East African savannas—the ones that chronicle how lions slaughter wildebeest, zebras, and buffalo and how leopards and cheetahs run down gazelles and antelopes—you may have noticed the strange contrast in the ability of the carnivores to kill their prey and to actually eat it. Big cats and hunting dogs are brilliant at chasing down and dispatching large herbivores, grasping the neck in their powerful jaws and severing their spinal cord or throttling them. However, when it comes to opening up the carcass and accessing its flesh, they struggle. They pull ineffectually at the corpse for hours, like a child trying to tear open a parcel. And even when the carnivores finally get through to the flesh beneath the skin, they struggle to bite off pieces of meat or offal or to break them up into pieces small enough to swallow and digest. They eventually leave the corpse half eaten and it ends up getting surrounded, sometimes for days, by a whole host of scavengers, who themselves take hours to tear off the food they need.

In contrast, if we return to the Dorobos, whom we met in the prologue, the secret of their successful theft, from a troupe of lions who had been mauling the wildebeest for the best part of a day, was that a single man was able to slice off a joint of meat in seconds. As we shall see in this chapter, one of the keys to the success of human hunter-gatherers is their hyperefficiency at food processing. And once again this human ability is down to the way people use power rather than strength to achieve their aims. Processing food is rather more complex than hammering or throwing, however, and to understand how people have become so good at cutting up animal flesh we first have to take a closer look at the mechanics of the fracture of soft biological tissues—how they break.

The reason that carnivores are so hopeless at feeding lies in the mismatch between the tools and methods they have available—their conical teeth and hinged jaws—and the mechanical properties of the tissues they have to deal with: skin, tendons, ligaments, and muscles. All of these materials are soft; skin, tendons, and ligaments are largely composed of the rubbery protein elastin, while muscle is made of flexible muscle cells. You might think they would be easy to cut up. However, though they are flexible, they are also strong, since they are reinforced by fibers of the rigid protein collagen. In skin, a random feltwork of collagen fibers limits how far it can stretch and causes skin to contract inward when it is stretched, a design that allows us to move freely while our skin hugs the body like spandex or Lycra. In tendons and ligaments the collagen fibers are oriented along their length to stiffen them, like the collagen fibers that make up the marbling in muscles and which transmit the forces generated by the muscle cells along to the tendons. The reinforcing collagen fibers in all of these tissues also toughens them, just as the stiff fibers within the plastic matrix toughens composite materials such as fiberglass. Breaking these tissues produces a rough fracture surface that takes a lot of energy to produce. The fibrous nature of these materials also confers a further advantage: once a fiber snaps, the soft material around it stretches, blunting the tip of a crack. This means that the force is shared evenly between the remaining fibers rather than being concentrated at the tip of the crack. It makes soft biological tissues almost impossible to tear.

Consequently, when a predator tries to cut through skin or break off a piece of flesh by biting into it and pulling a mouthful away, its pointed canines will puncture it, but, unlike the perforation lines in stamps, will hardly weaken the tissue at all. As the animal pulls, the holes will stretch along the line of pull, just as cling film does when you try to pull it apart with your bare hands. The material simply refuses to break. You might think that a predator could instead use its sharp claws rather than its teeth to tear through flesh. However, because the claws have a circular cross section, this would be even less successful than the predator's using its teeth, as studies on dinosaurs have shown.

Back in the 1970s, the American paleontologist John Ostrom claimed that the small therapod dinosaur *Deinonychus* could have used a huge curved claw on its hind feet (*deinonychus* is Greek for "terrible claw") to "disembowel" herbivorous dinosaurs with a single kick. A colleague of mine at the University of Manchester, Phil Manning, set up an experiment to test this idea. Working in collaboration with a TV company (dinosaur documentaries are ratings gold), his team made a hydraulically powered model of a *Deinonychus* leg and armed it with a reconstruction of the claw. They then set it up so that it kicked out at the belly of the corpse of a pig, the closest modern equivalent to the stomach of a dinosaur. The results demolished Ostrom's theory. The curved claws could not tear the skin at all. Instead, as they hit the skin, they hooked into it and were held fast; they did not act as knives but as crampons. *Deinonychus* would have used its huge claws to hold on to a herbivore so that the therapod could employ its sharp teeth to bite into its neck or other vulnerable parts of its body. The claws of the big cats of today do just the same thing when they are catching their prey, as you can see in those wildlife films; but they are just as ineffective at cutting through skin.

Lacking the ability to slice through the body tissue of a herbivore by clawing at it or biting into it, a modern-day predator such as a lion has to grip a part of the body with its incisors and pull as hard as it can to attempt to break through the skin by brute force and tear off some flesh. It may make repeated bites to try to weaken the tissue and shake its head from side to side to produce large dynamic forces. But even when it has finally succeeded in biting off a piece of flesh, a predator's battle is not over. Its highly cusped molars are almost as unsuited to breaking it up into smaller pieces as its canines; most predators end up swallowing huge chunks of meat that can take days to digest. This is one of the reasons why lions rest for so long between hunting expeditions.

In contrast, as the video of the Dorobos shows, human hunters can cut through meat with just a few well-aimed strokes of their knives. Of course, they use metal knives, so it is tempting to attribute their success to their exploitation of modern metallurgy. But hunter-gatherers were able to butcher animals efficiently using simple stone tools well before the advent of metals. Stone flakes can be even sharper than metal knives. And the abilities of humans to cut meat are not just due to the sharpness of our knives, but to the way in which we use them. As any part-time cook who has tried to carve the Christmas turkey and ham will have found to their cost, even if you use a sharp knife, cutting meat is not as easy as it looks. Success depends on the skilled actions of the carver.

The instinctive way to approach cutting through a material is to push the blade down directly into it, as we do when we chop brittle vegetables such as carrots, onions, or nuts. The blade of the knife acts as a narrow wedge, which produces a crack just in front of its tip, creating high tensile stresses that break the material in front of the blade and force the crack to run straight through the food. Unfortunately this technique will not work to cut up foodstuffs such as tomatoes or meat that are soft but tough. Because these foods deform so easily, instead of opening up a crack, the compression force at the tip of the blade merely squashes the material in front of it. Only when the knife has depressed the surface so much that it forms a steep valley that acts like a supersize crack does the food fail in tension at the tip of the crack. However, even after this initial failure, the fracture surface opens up again and the knife deforms the food even more. The process demands a large force, uses up huge amounts of energy, and creates a squashy mess.

To get over this difficulty, both the Dorobos and carvery chefs use a very different technique. They slide their knife back and forth across the surface of the meat while applying only a relatively small force downward. This slicing method cuts the food up far more efficiently and for a somewhat surprising reason. Applying the sideways (what engineers call shear) force on the surface of the food sets up tensile forces at forty-five degrees to it, forces that stretch and break the surface skin and open up a crack. Continuing to slide the knife back and forth in this crack carries on the process and deepens it. The action is efficient because almost all the energy is used to produce the two new fracture surfaces on either side of the knife; none is wasted deforming the meat. And you should be aware of just how effective the mechanism is because it causes that most painful yet least heroic wound: the paper cut; even floppy sheets of paper can cut readily through our skin if we accidentally slide the edge across our fingers. Using the slicing method greatly reduces the time and energy we need to cut through skin and sinews, to remove flesh, and to cut meat up into digestible slices. Moreover, it is easy to do, as we can grasp a knife firmly in our hands and slide it back and forth using our mobile arms as cranks. We use the muscles of the arm and shoulder to provide the necessary power, swinging our upper arm forward and back at the shoulder and extending and flexing our elbow.

As far as I am aware, no other animal uses such an efficient slicing technique. Once they have broken off a piece of meat, big cats can break it up to a limited extent by gnawing at it with their triangular molars, but our primate cousins have no equivalent way of cutting up meat. The chimpanzees that hunt and eat colobus monkeys simply tear these small creatures apart with their bare hands.

But though it is straightforward to use a knife, it is not so easy to find or make one in the first place. After all, early hominins sourced their hammer stones from dried riverbeds, where most of the pebbles had been rounded off by water erosion. We are fortunate, therefore, that 4 million years ago our hominin ancestors, the australopithecines, were already using tools, stones, that could be made into knives, and that they were already using these stones in such a way, as hammers, that they might accidentally break, creating slender flakes that they could use as knives. And to see how you can make a stone knife merely by hitting two stones against each other, we need first to understand something more about the fracture properties of stone.

Stone is so stiff that stone hammers crush more compliant objects, transferring kinetic energy into them to break them, while the stone itself remains intact. However, if two stones are hit together, the outcome will be very different. Though stone is stiff and hard, it is also extremely brittle because cracks can readily travel between the atoms and molecules. Consequently, if two stones collide, the energy will fracture one or both of them. In a crystalline material such as quartz, or in a sedimentary rock such as sandstone, the stone will fracture along an existing line of weakness. In an amorphous one such as flint or chert, in contrast, the stone will break along lines of stress that radiate around the point of impact. Knowing this, a skilled stone knapper can predict exactly how a particular stone will break.

Australopithecines may have found that their hammer stones broke if they accidentally hit the anvil stone when they were cracking nuts or when they were trying to crush bones. They may subsequently have noticed that if a flake of stone had been dislodged from the hammer, it would give it a more pointed surface, which would concentrate the impact of the hammer onto a smaller area of bone, crushing it more effectively. And they might also have noticed that the slender flakes that had sheared off the large hammer stone had razor-sharp edges that they could use to slice through skin and flesh. Particularly gifted individuals might even have realized that merely by hitting one stone with another they could break it up in a controlled manner; if they hit a large "core" stone near its edge, they could dislodge "flakes" of stone, with a thick end that they could hold in the hand, and a distal end with a sharp edge that was good for cutting. They could make a new sharp cutting tool—a stone knife.

We cannot be sure when the first australopithecines created the first stone knives, or when they first used them to slice open skin and saw through the meat. The first evidence that they were using stone knives, however, dates back some 3.4 million years. In 2010, a team of anthropologists led by Zeresenay Alemseged from the California Academy of Sciences found small grooves on the bones of herbivores at a site in Afar, Ethiopia. The evidence was equivocal since the grooves could have been produced in another way, but in 2015 a multinational team of anthropologists working at an only slightly later site, dating back 3.3 million years, at Lomekwi, Kenya, found evidence for an actual toolmaking industry. They found eight-inch-diameter (20 centimeter) stone anvils, four-inch-diameter (10 centimeter) cores, and two-inch-wide (5 centimeter) flakes. An australopithecine, *Kenyanthropus platyops* or possibly *A. afarensis*, had broken leaflike flakes from the cores, either by holding the core like a hammer and hitting it on an anvil, or by placing the core on the anvil and hitting it with a hammer stone. This technique clearly worked, but the results were variable: the flakes, from what the authors dubbed the Lomekwian industry, were rather large and crudely formed compared with later ones. Many cores had signs of damage well away from their edges, where a misplaced strike had failed to dislodge a flake. The poor craftsmanship was probably because it is quite difficult to accurately judge the distance of an object relative to our body. As anyone who has tried to hang a picture on a wall can attest, it is all too easy to miss your nail and hit your thumb instead or strike it a glancing blow, bending the nail out of shape. And it is extremely tricky to strike a log of wood with a long splitting maul in the exact position needed to split it. Typically the longer the swing and the more powerful the blow, the less accurate it is.

Fortunately, by 2.9 million years ago, hominins had greatly improved their flaking technique. In 2023, another multinational team working at a site in Nyayanga, Kenya, found the earliest evidence yet of the stone-knapping technology that was to dominate stone toolmaking for over a million years, the Oldowan industry. Rather than using an anvil, the australopithecines, probably members of the genus *Paranthropus*, had learned how to hold the core within one hand, cradling it between thumb and palm, and to hit it with a hammer stone held in the other hand. Of course it is not possible to apply a powerful blow using this technique since one cannot make use of the rotation of our hips or shoulders to generate a two-stage sling action. However, given the brittleness of stone, this is not a problem; typically it only takes one to two joules of energy to split a core, far less than the eight to ten joules we can generate just by swinging our arms. The great advantage is that this technique can be much more accurate, since the proprioceptors in our bodies tell us precisely the relative positions of our limbs; after all, our hands never miss each other when we clap. Another advantage of the technique was that the toolmaker could hold both the hammer and target stones at just the right angles so that the hammer could hit at the correct spot, and in just the right direction, to dislodge a flake of a desired shape and size.

Since they were produced by a more controlled technique, the Oldowan tools were a great advance on the earlier Lomekwian technology and made food processing more effective. The australopithecines made heavy handheld "choppers" around four inches (10 centimeters) in diameter by hitting a target stone in alternate directions along the sides to produce an obtuse cutting edge. They could hold these choppers in a power grip within their palm and use them to crack open bones. Meanwhile, they could hold the one-inch (2.5-centimeter)-diameter flakes that had split off from the choppers, grasping the blunt end in a precision grip between their fingers and use them to slice open skin and cut through flesh. Recent X-ray investigations of the fingers of australopithecines revealed that the trabecular bone reinforcing the walls of their fingers was well-developed, showing that they must have spent much time in activities that demanded a strong precision grip.

The australopithecines even learned how to make customized flakes for particular purposes. They could slide sharp flakes back and forth between the skin and flesh of corpses to gradually separate the two tissues and remove the skin. They could hold specially blunted flakes with the blade at right angles to the motion and use them as scrapers to remove fat from the inside of the skin and hair from its outside to make durable animal hides. They could even produce flakes whose edges came together into a sharp point and use them to cut notches in bone, or to rotate them back and forth to drill holes.

By 2.5 million years ago the australopithecines had become the first true manufacturers, people who not only used tools but created them. From then onward most of the advances made by hominins would be made not in the form of their bodies—after all, the late australopithecines and the early members of our own genus, *Homo*, looked very much like us—but in the quality of the tools they made and in the sophistication of the techniques they used to manufacture and use them.

There would have been strong selection pressures on individuals to improve their fine motor skills and develop better mechanical insight. But this alone would not have been enough to drastically speed up technological improvements. Any genetic changes in brain structure would take many generations to spread through the population. A far quicker way of spreading technological change would be for individual hominins to improve their skills via a social rather than genetic mechanism—to simply copy the techniques used by the more talented members of their group. As the psychologist and professor of evolutionary biology at Harvard University Joseph Henrich has shown, this would lead hominins to become more social beings; they would not only become better at understanding the physical world, but also other individuals' behavior; they would be better judges of who would be the best to copy. So the selection pressures that would lead to rapid improvements in engineering, toolmaking, and material culture would also lead to the evolution of a more complex culture and language as hominins formed stronger social bonds within their groups.

Two further factors could also have accelerated how fast tools could "evolve." The insight of a particularly talented individual could allow the group to make a large technological leap to produce a wholly new tool, in a way that seldom occurs in natural selection; most "hopeful monsters" in nature are quickly eliminated. A single person could also adapt or use an existing tool for a completely new purpose. Human technology could begin to develop at an increasingly frantic rate.

But there were costs. For a start, greater technical and social skills would have demanded greater computational power, so it is not surprising that by 2.5 million years ago the australopithecines had developed brains that were around 25 percent larger than those of chimpanzees. Since the brain uses such a large amount of energy, this would be an additional metabolic burden. And the requirement for a powerful, versatile brain also led to another burden: a long, helpless childhood. Apes, which have to learn how to survive in their dangerous treetop environment, already take a long time to attain the competence they need to reach an independent adulthood; young orangutans require seven years, for instance, to perfect the building of their nests. Hominins would have had much more to learn about how to navigate their tool-rich environment, so it would take them even longer to become productive members of the group.

Hominin infants would also be far more helpless than our primate cousins and would have had to be carried around for months or years. To retain the huge potential to learn that they required, their infant brains would have to be developmentally extremely plastic; they have to be modified to produce and use tools, and to fit in with the material culture of their group. To achieve the necessary plasticity, they would have to be born with a brain that is at an early stage of development. At birth, only 1.6 percent of the neurons of modern humans are myelinated, compared with 15 percent in chimpanzees. Since the speed of nerve conduction in unmyelinated neurons is much lower, young humans are necessarily less well coordinated than young apes. Humans take much longer to develop the motor skills they need to crawl, stand up, walk, and run, and to develop basic craft skills. In *Australopithecus africanus,* childhood seems to have already stretched to twelve to fourteen years. With their complex stone tools and developing social skills, and with their longer childhoods, by 2.5 million years ago the australopithecines were becoming more like us mentally as well as physically. And with the emergence of *Homo erectus* just over 2 million years ago, there came a creature that we might almost regard as human.

Chapter 5

THE TRANSFORMATIONAL TOOL

The environment in which the early hominins were evolving did not stay the same. East Africa continued to dry out, and tree cover fell further. Hominins had little option on where they would live. Two million years ago a new species of hominin, *Homo erectus*, emerged that was better adapted physically to life on the open plains, becoming the first totally terrestrial hominin, and was able to broaden its ecological niche so that it was not just a gatherer of plants and scavenger of animal corpses, but an apex predator.

Its most important physical adaptations were in the size of its limbs and shape of its body. *Homo erectus* developed longer and more powerful legs than australopithecines, with greatly enlarged gluteus maximus muscles; and it had broader shoulders but more slender arms. This would have enabled it not only to walk and jog, but to sprint. It used the power developed by the muscles acting around its hips and knees—the quadriceps, hamstrings, and particularly its greatly enlarged gluteus muscles—to swing its legs back and forth during the flight phases between footfalls. And its broader shoulders enabled it to swing its arms more effectively in the opposite direction to its legs, counterbalancing the torque they produced, and keeping the body balanced. This enabled *Homo erectus* to run as fast as modern humans and make up some of the speed disadvantage hominins had always labored under relative to other terrestrial mammals. *Homo erectus* would therefore have been a more capable predator and less vulnerable prey than its ancestors.

This form of bipedal running is also a surprisingly efficient form of locomotion. We can store and release energy in our Achilles tendons and the arch of our feet, enabling us to bounce along as we run, and we can also store the energy needed to swing our arms and legs in elastic structures in our back and in the tendons at the ends of our hamstrings and quadriceps muscles, reducing the amount of energy we need to expend. Some anthropologists have suggested that *Homo erectus* might even have been an endurance predator, like a hunting dog, chasing game animals until they became exhausted or overheated in the sun.

But a more important benefit was that, with its broader shoulders and enlarged gluteus maximus muscles, *Homo erectus* would also have been able to deliver more powerful blows than australopithecines, and its straight fingers would have improved its grasp of weapons such as hammer stones, clubs, and spears. It would have been a far more formidable creature, capable of killing large animals whenever it encountered them. It would have been better at using Oldowan tools than its australopithecine cousins. It could have wielded hammer stones and wooden clubs to break hard, brittle materials such as nutshell, mollusk shell, and bone, giving it access to seeds, seafood, bone marrow, and brains. And it could have used stone flakes to chop soft, brittle foodstuffs such as fruit, vegetables, roots, and nuts, and to slice through soft, tough materials such as skin, tendon, and muscle. Together the two sets of Oldowan tools would have made a fine set of kitchenware and cutlery.

On its own, however, *Homo erectus*'s modified physique would not have been enough for it to survive in the savanna because the redesign of its body did have one main disadvantage. With its muscles redistributed from its arms to its legs, and with a more laterally pointing shoulder joint, *Homo erectus* would have had weaker arms and would have been far less able to pull itself upward into the canopies of the few trees that remained in the savanna. This would have given it several problems. First, it would have reduced its ability to pick fruit and nuts, and to reach and break off the slender branches of trees that it would have needed to make wooden tools such as digging sticks, spears, and clubs. It would also have been unable to make a sleeping nest in the tree canopy; even if it could have climbed up a tree, it would not have been strong enough to break off large branches and build itself a nest. Stranded on the ground with fewer means of obtaining food, lacking a nest to sleep in or wooden weapons for defense, it would have starved and been easy prey for large carnivores.

The root cause of these problems is that wood has exceptional mechanical properties. Australopithecines could use hammer stones to crack open nutshells and animal bones because though these materials are stiff, they are also brittle. The australopithecines could use stone choppers to break up vegetables because they are both soft and brittle. And they could use sharp flakes to slice through skin, tendon, and meat because, although these are tough, they are also soft. Wood, however, is both stiff and exceptionally tough, especially across the grain, so it is hard to snap or cut up. Hitting a wooden branch with a hammer or chopper would hardly dent it. Trying to slice it across while holding a flimsy flake between the fingertips would merely engrave a narrow groove on its surface. And trying to break it by bending it would simply cause it to buckle, or to split along its length after it had broken just half way across, in a pattern of failure known as greenstick fracture. It would be almost impossible to snap in two.

Modern chimpanzees make their spears and digging sticks by tearing small branches off at their junctions with larger branches and sharpening the tip with their teeth. Consequently, their wooden tools are barely modified sticks. *Homo erectus* needed a new stone percussion tool to produce larger, more sophisticated wooden tools. It needed a stone tool that was sharp enough to concentrate force; heavy enough to store enough kinetic energy to cut through a branch; and rounded enough that it could be held firmly in a power grip and not damage the user's palm on impact. Yet the tool still needed to perform all the functions of the Oldowan tool set: to slice through the flesh, skin the carcass, scrape the hide, and break open the skull and long bones. *Homo erectus* could have amassed a whole collection of tools to do this—choppers, knives, scrapers, and burins—but it would have been almost impossible to carry them all around. Instead they would have had to haul the whole carcass back to their camp to process it, which would involve much more work than processing the carcass at the scene and bringing back just the best parts.

Combining all the tools into one would therefore be incredibly useful. I'll certainly always remember the day my life was transformed simply by the acquisition of a multipurpose tool. I was returning from a plant biomechanics conference in the Black Forest spa town of Badenweiler with my PhD student Bobby Mickovski, and we had half an hour to kill at Mulhouse airport. In a shop in the departures lounge they had an offer on Swiss Army knives. On a whim I bought the Explorer and have never regretted my choice. The collection of features, from knives, scissors, screwdrivers, tweezers, can opener, bottle opener, corkscrew, and magnifying glass gave me a whole new feeling of omnipotence. Never again would be unable to open a parcel; never again would I struggle to change a fuse; never again would I have to look on, frustrated, at an uncorked bottle of wine. I was transmogrified in a flash from an effete intellectual into a man of action.

Homo erectus had a more difficult problem since it had to design and make its own multipurpose tool. That tool was the hand ax, which combines the virtues of ax, hammer, and flake all in one. A typical paleolithic hand ax is much larger than a flake and more finely modeled. It has a flattened teardrop shape, being around four inches (10 centimeters) long, two inches (5 centimeters) wide, and four-tenths of an inch (1 centimeter) thick, and weighs around a pound (0.5 kilogram), similar to a hammer stone. It has a rounded back that fits snugly into the palm, tapers to a point at the other end, and has its two sides flaked into sharp blades along each edge. It can do everything that a flake can. The blades along either edge can chop through vegetables, be swept from side to side along the join between the skin and the body to shear off a hide, or dragged across soft material to slice through tendon and muscle. It can do everything that a hammer stone can; being heavy and having a sharp point, it can give a prey animal or a human enemy an even more deadly blow than a simple hammer stone. And since it is both heavy and sharp and has a broad smooth base, it can also be used in an entirely new way that combines the action of a knife and a hammer—as an ax that can cut through wood.

But even a hand ax cannot cut through a branch if you hit it head on. Wood is so tough and stiff along the grain that even the heaviest blow would hardly penetrate at all; the ax head would end up being stuck in the wood. It must be used in a way that exploits the weakness of wood across the grain. To make an impression the user of an ax needs to sweep it at an oblique angle up or down the stem of a sapling or a branch of a tree, so that it splits it as well as breaks it across, producing a deep gash. The user can subsequently repeat the procedure cutting alternately up and down and all the way round the bough, eventually severing it, and producing a pencil-shaped point, like that produced by a foraging beaver.

Armed with a hand ax, enthusiasts of primitive technology can cut down a sapling or sever a branch of a larger tree in minutes, and they can then use the ax to carve it into shapes. They can either apply angled blows along the grain to shave off slivers of bark and wood, just as a traditional woodsman uses an adze, or slide the blade of the ax along the wood, using it like a drawknife or a modern plane. With practice a hominin, using just a single hand ax, could have formed a range of large, finely shaped wooden tools, from stout digging sticks to full-size throwing spears. Wood residues found on the blades of hand axes show that these tools were indeed being used for carpentry over 1.6 million years ago. The invention and use of the hand ax enabled *Homo erectus* to produce a whole range of tools made from wood, antler, or bone. Having larger digging sticks would enable it to dig up deeper roots, for instance, while having longer, heavier spears would enable it to kill larger mammals, even at a distance.

The development of the hand ax marks a major step in human evolution. For a start, the manufacturing was far more complex. Whereas breaking off a flake is performed in an instant, the ax needs to be gradually transformed. The maker has to gradually free it from inside a stone nodule, like Michelangelo's *David* from its block of marble. First they had to choose a nodule of approximately the correct size and shape, then shave flakes off the side of both blades by striking them a glancing blow with a hammer stone, gradually creating the two cutting edges and the point, while leaving the base smooth. Hand ax makers plainly knew in advance what particular purposes they wanted to use the axes for because they customized them. The most common variant was the chopper, which had a straight blade at the front end rather than a point and was better at chopping vegetables or chiseling wood than slicing meat.

The hand ax was also the first tool that was not just made and used on the spot to satisfy an immediate need, such as to kill an animal or cut up food, but was made for future use; it was the first tool that was made to make other tools. This shows the developing ability of hominins to plan ahead and aim for more and more distant goals, and it demonstrates *Homo erectus*'s ability to delay gratification, which is such a mark of human thought.

Using its hand axes, *Homo erectus* could also transform many other aspects of its life. It could construct large, semipermanent huts, which are essentially inverted sleeping nests. Like the huts of modern-hunter gatherers such as the Hadza, *Homo erectus* would have built them by cutting down a number of tree saplings and inserting them trunk down in the soil to make a circle of the desired size. They could then bend the stems over and weave them together at the top, before covering the structure with a roof and walls made of grasses or leaves. Though woven wooden huts are flimsy, they would have helped *Homo erectus* keep warm during the cool savanna nights by protecting them from the cold night sky and sheltering them from the wind, which could raise the effective temperature inside by up to five and half degrees Fahrenheit (three degrees Centigrade). They would have been able for the first time to sleep not alone in nests, but together in family groups, which would also have kept them even warmer and allowed them to lose their body hair. It would also have helped protect them from predators and would have had an important impact on the ways in which these hominins socialized. Having family groups sleeping in the same hut would have promoted monogamy and further improved their communication skills. Tantalizing evidence of just such a hut, in the form of a ring of stones thirteen feet (4 meters) in diameter, has been found in a 1.8-million-year-old site at Olduvai Gorge in Tanzania, suggesting that even the earliest *Homo erectus* were competent builders.

Once settled in semipermanent camps, *Homo erectus* could also use the twigs, wood shavings, and larger branches that it had hacked off trees or collected from beneath them to produce a wholly new form of power—heat. Wood is an unusual biological material in many ways. For a start, as it dries out, its mechanical properties actually improve: it gets stiffer without getting any weaker or more brittle. And as its water content falls, wood also becomes much more flammable. Heating it up can release the chemical energy in its cellulose fibers, hemicellulose matrix, and lignin cross-links. The process is difficult to start, however. At temperatures above 212°F (100°C) all the water is driven off; at 300°F (150°C) the hemicelluloses start to crystallize, which makes the wood harder and more brittle. Only when the temperature reaches around 390°F (200°C) do the large molecules start to break down, a process called pyrolysis, releasing energy and raising the temperature further. Eventually, this causes the smaller molecules to vaporize and, above 570°F (300°C), to react with oxygen in the air to form a flame and release much greater amounts of energy. As the process continues, the fire gets even hotter until all that is left of the wood is charcoal, the carbon skeleton of the tree, which reacts with oxygen at its surface to glow red-hot at temperatures exceeding 840°F (450°C).

Fire is usually seen as destructive, but it can also be used for constructive ends. A fire can keep people warm on cool savanna evenings before they retire to their sleeping huts. Burning wood produces around fifteen to twenty megajoules of energy per kilogram, so a large fire, consuming ten kilograms of wood over four hours, would produce something like ten to twelve kilowatts of power—enough to keep the people huddling around it warm, as we can all remember from our days singing around campfires. Fires also produce light, so they would have enabled the hominins to carry on performing domestic duties, and to carry on interacting and communicating among themselves well into the night, further developing their social bonds and language skills.

But though *Homo erectus* would have been able to harvest and cut up enough wood to burn using their hand axes, they would have found it far harder to set light to it. Fortunately, in dry savannas forest fires are fairly common, so it would have been relatively easy to obtain a fire that was already lit. And once obtained, it is quite easy to keep a fire going, and even to transport it. Modern groups of Australian nomadic aborigines, for instance, carry smoldering logs with them, which they can readily fan into flame when they reach their new camp. *Homo erectus* could have done much the same. We cannot know for sure exactly when hominins first controlled and made use of fire, especially considering that the burnt remnants of campfires can leave little trace even a few days after they have been extinguished. However, anthropologists agree that definitive evidence of controlled fires date back at least to eight hundred thousand years ago, to sites inhabited by *Homo erectus*, which was still the dominant hominin species on the planet.

Having gained control of fire, *Homo erectus* likely also used the heat generated by their wood fires to enhance and extend their diets, by cooking their food. Heating meat to temperatures over 160°F (80°C) breaks down the collagen fibers in the marbling, and heating plant material breaks down the pectins that glue the cells together and weakens the hemicelluloses in the cell walls. Cooking food therefore makes it much softer and easier to chew. Consequently modern hunter-gatherers spend far less time eating than chimpanzees, around an hour each day, compared with five hours for chimps. And rather than needing flat platelike teeth to break up raw food as the australopithecines did, *Homo erectus* could break up its cooked food with its much smaller, high-crowned teeth, teeth that closely resemble our own. They could also dispense with the powerful chewing muscles of australopithecines, and the saggital crown along the top of their skulls, on which they inserted. The skulls of *Homo erectus*, with small teeth, short muzzle. and higher, smoother crown, were far more like ours.

Soft, cooked foods are also far easier to digest; we can extract around 80 percent of their energy, compared with only 60 percent of that in raw food, and it requires around 12 percent less energy to digest. Consequently, *Homo erectus* was able to shorten its gut, dispensing with the potbelly of apes and australopithecines, and to divert more energy to developing and supporting a larger brain. With a volume of over eight hundred milliliters, the brains of early *Homo erectus* were 50 percent larger than those of australopithecines and twice as large as the brains of chimpanzees.

The impact of all these changes enabled the species to become more sociable, improve its diet, and revolutionize its communication skills. *Homo erectus* would have been able to live in larger groups, bringing their food back to semipermanent camps and sharing the food with other members of the group. And in turn this set off a further runaway series of behavioral and technological changes that involved a further 50 percent increase in brain size. Between 2 million and eight hundred thousand years ago, their brain volume rose from eight hundred to twelve hundred milliliters. The tools they created also attest to their rising technical intelligence. Though *Homo erectus* continued to use hand axes throughout their history, later on they revolutionized their manufacture to improve their cutting efficiency. The original technique, using stone hammers to shave a series of flakes along each side, produced curved fracture surfaces, so the blades along the sides of the ax were wavy and the hand axes had to be rather thick. Though good as hammers, they therefore performed relatively poorly as knives. Later *Homo erectus* devised a more sophisticated manufacturing technique. Rather than attempt to produce the cutting edges in one go, they did something totally counterintuitive: they cut a platform all the way around the core. Only when that was completed did they strike the platform along its edges with a wooden baton, removing first one side of the blade and then the other to produce long, almost-straight cutting edges and a much more slender and sharper ax.

Being able to deal with a whole array of materials, from the softest and most brittle to the hardest and toughest, with just a single perfectly manufactured power tool enabled *Homo erectus* to survive for nearly 2 million years. And it would give birth to a range of daughter species that would in turn make and use a much-wider range of more specialized tools and weapons.

Chapter 6

MIXING AND MATCHING

For all its many virtues, my Swiss Army knife is compromised; by being a jack-of-all-trades it is master of none. It is small, tricky to hold firmly and manipulate, and has flimsy blades. So, though I always have it in my pocket, at home I invariably use one of my hundreds of single-purpose tools. I have a house full of kitchen utensils, cutlery, stationery, string, and tape; a garage full of screwdrivers, spanners, drills, saws, chisels, and sandpapers; and a shed full of forks, spades, pruning shears, and hoes. I only use my Swiss Army knife away from home or if I am at the bottom of the garden and have none of my purpose-made tools to hand.

Homo erectus must have found itself in a similar situation. The hand ax would have proved clumsier at performing cutting and scraping tasks than a purpose-made flake, and it would not have had such a sharp cutting edge. It would also have been awkward to use as an ax, since an impact on solid branch wood would jar the user's hand. *Homo erectus* could hurt itself, too, by accidentally hitting their hand against the target, and splinters of wood could force their way through their skin. A recent experimental study carried out by Dominic Coe and colleagues at the University of Liverpool has shown that modern humans limit the power they apply and the energy they build up when they use a handheld tool to cut up wood so they help minimize these dangers.

57

Over the Middle Paleolithic, therefore, new emerging species of humans, *H. heidelbergensis*, *H. ergaster*, and our close cousin, Neanderthal man, *H. neanderthalensis*, developed a series of new technologies. They improved the ways in which they made stone tools and learned how to combine stone, wood, and other materials to produce a whole new range of composite tools that were more effective. This design revolution would set humans on the road to be the most efficient animal of all at obtaining and processing food.

Around five hundred thousand years ago humans developed the platform technique, first pioneered by *Homo erectus*, to create large numbers of precisely shaped flakes in what is known as prepared-core technology. In the Levallois technique, first identified and named after the Parisian suburb of Levallois-Perret, the knapper would first choose a large stone nodule and flake off its sides to produce what was essentially a thicker version of a hand ax blank—a core. They would then peck off a series of flakes around the top surface so that it resembled one side of a hand ax. Finally, they applied a sharp blow just below the top edge of the core, splitting off the whole upper dome and producing a large slender flake with a flat lower surface and a razor-sharp edge.

In the Mousterian technology, which was developed slightly later, knappers used a similar technique to produce a larger number of smaller flakes from a single nodule. The knapper shaped the core in much the same way as in Levallois technology, but rather than removing a single large flake off the top, they knapped the prepared core in the other direction, removing material all around the sides to produce a whole series of flakes. Not only did this technique create a greater number of tools, but the shape of each flake could be finessed by altering the shape and proportions of the prepared core. Toward the end of the Middle Paleolithic, for instance, around 120,000 years ago, knappers started to produce smaller, deeper conical cores, from which they could flake off huge numbers of fine, narrow, tapered blades, which were much longer than they were wide. With so many flakes being produced, the craftspeople could customize them for particular purposes, using a technique known as retouching. They could press close to the periphery of the flake with a bone tool to snap off one of its sharp edges, leaving it with a blunter side that could be used for scraping tasks. The tip of a blade could also be shaped, removing the edge at one side to leave a point. This produced a tool known as a burin. which they could use to engrave the surface of wood or bone or rotate in their fingers to act as a borer or simple drill. Alternatively, they could remove one whole side of a blade to produce a tool with a sharp edge along one side and a blunt back, which a person could hold safely and press on with their fingers to use as a knife.

These new tools would have been far more efficient at their jobs than the generalized hand ax. But they would all have had one major disadvantage; as they got smaller, they would have been far more difficult to hold and far more fiddly to use. The hominin would have had to grasp a flake in a precision grip between their fingertips, which would have limited the force they could apply, tire out their finger muscles, and make the tool far more dangerous to use. These developments in blade technology therefore went hand in hand with another technological revolution: the development of hafting. Instead of holding the stone tools between their fingers, humans started to find ways to graft them onto handles made of wood, antler, or bone. Hafting a blade has many advantages, as anyone who has had to use a razor to shave or a scalpel to carry out dissections will appreciate. You can keep your fingers well away from the cutting edge, making it easier to see and guide the blade, and making it safer to use. The biggest plus is that you can hold the handle in a power grip within the palm like a toothbrush, making it easier to control, and enabling you to apply a far greater force without tiring out your finger muscles. You can work far faster if your knife or scraper has a handle, use less energy, and suffer from far less muscle fatigue.

Hafting a blade is no easy matter, however. You first have to make a blade with an extended rear end, or tang, which you can insert into a handle. It is fairly easy to cut grooves into a piece of wood or antler to do this, but cutting the groove accurately to achieve a snug fit is extremely tricky. Even if you do manage it and can insert a blade into a handle, then it can come out again as well! To attach blades securely, people had to create yet another totally new technology—glue. They started to use a wide range of biological materials: the resin of coniferous trees and birches; the collagen from cartilage, ligaments, and tendons; and beeswax. They heated their glue above a fire to liquefy it so that it could flow over the joint and into the gaps between the two elements before the glue solidified and hardened as it cooled. They were using the power of heat for the first time in manufacturing.

One factor that must have helped them do this was a newfound ability not only to use fires, but to start them. They found that striking a blunt flint tool against a lump of iron pyrites liberated enough heat energy from the collision to produce sparks that could ignite tiny shavings of wood or flakes of fungus. This technique produces a characteristic wear pattern at the tip of the tool, so anthropologists can tell if a flake has been used for this purpose. The evidence shows that people could have been lighting fires up to 160,000 years ago, and not only did our own species, *Homo sapiens*, do this, but Neanderthals as well. In Africa and Australia *Homo sapiens* also used fires to bake the local stone silcrete, altering its structure and making it easier to knap. The pace of invention was speeding up as the different technologies acted synergistically, advances in one technology opening up opportunities in other fields.

Along with hafting slender blades onto handles to make slicing and scouring tools, the Middle Paleolithic saw humans for the first time producing and using stone heads on percussion tools, most notably those weapons most beloved of caveman films, stone-tipped spears.

But though stone points look impressive, it is not immediately apparent what advantage they would have given to Stone Age hunters. Early hominins must have been using spears for over a million years, and all the evidence demonstrates that they would have been perfectly good weapons. The Clacton spear (actually just the tip was found) dates to around 450,000 years ago and consists of a length of carved yew wood that tapers to a sharp point. The Schöningen spears, which date to 300,000 years ago, are complete versions of the same type of weapon and taper to sharp points at both ends.

You might not think that wooden tips would be hard enough to penetrate the hides and flesh of large herbivores, yet the ballistic tests that archeologists have performed have shown that they perform just as well as stone points. And the archeologists who discovered the Schöningen spears found them along with the skeletons of horses that the hominins must have killed using the spears. Some of the horses even had marks that had been made from wooden spear points. Tests that archeologists have carried out on replicas of these beautifully carved spears have shown that javelin throwers can accurately hit targets some sixty to one hundred feet (20 to 30 meters) away. The spears can be accelerated to speeds of up to 40 mph (18 meters per second), delivering seventy joules of kinetic energy, easily enough to penetrate the thickest of animal hides.

It seems unlikely that stone points would have improved the performance of such throwing spears. They would also have taken far longer to construct, and it would have been difficult and time-consuming to glue the stone tip to the wooden haft. Moreover, the stone tip would have been far more likely to break off on impact, though this may have increased the size of the wound in the prey animal and caused it to lose more blood. The benefit of hafting a stone point onto a spear may instead have been that it would have enabled people to make a dual-purpose weapon. Having a stone tip at the front of a short spear would draw its center of gravity forward, stabilizing it in flight. It could be used both as a handheld stabbing weapon and as a javelin and so be used for both defense and offense.

By the early Upper Paleolithic, fifty thousand years ago, therefore, humans had invented ways to construct and use a wide variety of complex tools. They were employing not only the whole range of tools that transmitted mechanical energy to the object being worked—hammering, slicing, and scraping tools—but were also modifying the materials with heat from their fires. But the complexity and difficulty of the tasks they had to perform to manufacture these tools had magnified manyfold. It would have taken many stages to make each part of one of their novel composite tools, and they would have had to conclude the manufacturing with a final assembly. And they had to do all this even before they could actually use their tools. This would have demanded a prodigious increase in their mental capacity, a demand that drove both modern humans and Neanderthals to increase their brain volume by a further 20 percent to reach around fourteen hundred milliliters by eighty thousand years ago. Humans would have needed a longer childhood and adolescence, during which they could learn everything they needed to know to survive and become skilled engineers. And they would have had more time and mental energy to devise less immediately useful activities: grinding up pigments to paint geometric patterns and wild animals; constructing and playing musical instruments; and simply chattering to one another. It would have made them even more like us.

But Paleolithic people's ability to transmit their power was still limited. They would not have been able to throw their stones or spears far and would only have been able to deal with the smallest saplings. And in the many parts of the world where there was no iron pyrites, they would still not have been able to light their own fires. There were still probably only one hundred thousand people on the whole planet. To dominate the world, humans had to come up with yet more technical innovations.

Chapter 7

TYING IT ALL TOGETHER

In her delightful childhood memoir, *Period Piece*, Darwin's grand-daughter Gwen Raverat bemoaned the fussiness of Victorian girls' clothes but waxed lyrical about one advantage they had over their twentieth-century counterparts—pockets. These were apparently so voluminous that she could carry about with her much more than the single Swiss Army knife that I can fit into mine; as well as scissors, one of her key items was string. In my 1960s suburban childhood, girls' large pockets had long gone, but string was still a vital part of our play. Girls would use it in their constructive and cooperative game of cat's cradle. Two girls would stand facing each other holding a long loop of string between them, fashioning it into bewildering patterns with obscure movements of their fingers. Boys predictably used string for a more destructive and competitive game, conkers. We collected the nuts of horse chestnuts, drilled holes through them, and attached them to the ends of pieces of string. One player dangled their conker while the other swung theirs, aiming to hit their opponent's conker. The positions were then reversed, and the game went on until one conker had been smashed to pieces.

This use in our games was just one of the many purposes to which string has been put in its long history; humans have been making and using string for over eighty thousand years. It has proved its worth in far more ways than in just playing games or tying things together. It also allowed humans to concentrate their muscle power even more, to hold their tools together, throw projectiles farther, to shape artifacts more precisely, and to start fires. It helped humans behave in a truly modern way and paved the way for the machinery we use in the industrialized world today.

The first material that people used as string was probably sinew or tendon, which has the advantage that it actually evolved to resist tensile forces. Consisting of bundles of stiff collagen fibers within a soft elastin matrix, tendon is both strong and notch resistant. It has its downsides, though. Its stiffness falls if it gets wet; it is prone to rot; and because the fibers are all parallel to its length, it kinks badly if bent, so it is poorly suited to tying knots. Later in the Paleolithic, therefore, people started to develop true string, which is composed of large numbers of narrow fibers, which they could harvest from several different sources. They could make it using cellulose fibers, taken from the stems of flax or hemp plants or the seed heads of cotton; or they could use keratin fibers, taken from human hair or the fleeces of sheep and other mammals. Both types of fiber have far better resistance to moisture than sinew, but have the disadvantage that, because each fiber is short, they need to be manipulated to be joined together. The solution is to twist them together along their length into helical strands, so that when the two ends of the strand are pulled, they tighten up and the friction between them pulls the fibers together.

String is particularly strong because even if a few fibers break, the tension they were holding quickly transfers into the adjacent fibers. String is also easy to bend because the fibers simply slip past one another and there is no kinking. String is therefore ideally suited to tie items together, and it was no doubt an integral part of the first clothes, being pulled through holes in animal skins to join different parts together and to give shape to leather garments. But people soon found that string had many other uses. For instance, it could help solve a problem that had bedeviled attempts to produce the hafted tools that people needed most urgently: axes.

We saw in the last chapter that from the Middle Paleolithic onward people were gluing their fine stone flakes onto handles made of wood, bone, or antler. This works fairly well for tools such as knives that are subjected to modest static forces, or for the points of projectiles that are loaded axially, so that the forces push them farther into their handle. However, as any DIY enthusiast knows, glued joints can form a weak point in any structure. The difference between the stiffness of the substrate and that of the glue concentrates stresses, and glued joints are all too prone to fail at low loads, either due to a failure of adhesion between the glue and the object, or to a failure of cohesion within the glue itself. Glued joints subjected to large dynamic loads simply break. This presents a particular problem in the design of axes and hammers. An ax head or hammer stone attached to the end of a wooden or antler handle should be able to deliver a far harder blow than one held in the hand, but these blows develop huge dynamic forces on impact, which rapidly destroy any glued joint.

One way of getting over this problem is to jam the stone ax head into a hole in the wooden handle, so that it is held fast. This would certainly hold a slender ax head securely, but it would seriously weaken the wooden handle. Since wood is far weaker across the grain than along it, repeated blows could quickly push the ax head into its mounting, acting as a wedge and splitting the two sides apart. Possibly because of this problem, hafted axes appeared late in human history, only at the very end of the last ice age, some ten to fifteen thousand years ago, when deciduous forests were spreading across northern temperate regions. People needed to deal with ever-larger trees and came up with a range of engineering solutions to produce cutting tools that could fell larger trees and shape wood. One method was to use a handle with a swelling at its far end, and to cut a broad hole through it, so that it held the ax head by the top and bottom surface rather than by its sides.

This ensured that a blow would not wedge the handle apart. Another method, commonly used in mainland Europe, was to wedge the ax head indirectly into the handle via a shock-absorbing plug of antler. A third method, seen in the tomahawks used by the Iroquois of northeast America, was to reinforce the joint by wrapping string all around it.

But people also came up with a totally different design of woodcutting tool. In the adze, the blade is held at right angles to the handle rather than parallel as in the ax. Adzes have two main advantages. First, users can swing an adze directly in front of the body rather than from the side as in axes, so they can aim the blade more precisely at a tree trunk, enabling people to fell trees more easily. Second, the adze head can be secured without fear of splitting the handle, by cleverly making use of a unique design feature of trees; they reinforce their branch junctions by wrapping the wood fibers around each other, a design that prevents the branch from being torn off. People could carve a handle for an adze by removing the base of a branch and a piece of the trunk it was attached to. They then cut a notch at the end of the piece of trunk and tied and glued the adze head into it. All the impact stress on the adze head would be transmitted up into the handle without dislodging it.

The benefits of hafting an ax or adze are huge. The recent tests carried out by anthropologists from Liverpool University have shown that hafted tools can be swung far faster and be more effective at performing simple chopping tasks than handheld ones, even when the subjects are kneeling to perform the task. By the time of impact the users had accelerated the head of hafted axes to three times the speed of handheld ones and so delivered ten times the energy to the target, around ten joules. When hafted axes are used in the field, with the users standing up, tree fellers can build up much larger amounts of energy in each swing, so that they can cut down even full-grown trees. In modern lumberjack contests, for instance, competitors using a two-handed technique and employing muscles from all around their body can generate up to four hundred joules in the heads of their two-pound (1-kilogram) axes.

Even before people had developed hafted axes, they had also found ways to use string to propel projectile weapons to unprecedented distances. Since you can throw stones much faster and farther than large rocks, you might think that you could throw small pebbles even farther and throw lightweight arrows much farther than heavy javelins. Unfortunately, this is not the case because of the limitations of our muscles. The faster a muscle contracts, the less force it can produce and the less energy a single contraction can generate. Consequently, if you attempt to throw a lightweight projectile, your muscles can only accelerate your hand to a slightly higher speed than if you were throwing a heavier one. Moreover, the energy you did generate would mostly get used to accelerate your hand and arm—not the projectile.

This makes it all the more surprising that archeologists, led by Laure Metz and Ludovic Slimak of the French National Centre for Scientific Research, have found fifty-four-thousand-year-old projectile heads with a diameter of only half an inch (10–15 millimeters). These must have been mounted on sticks that were the size of modern arrows, with shafts that were around two-fifths of an inch (10 millimeters) in diameter and less than a yard (90 centimeters) long. These arrows must have weighed around two ounces (55 grams). Some archeologists have claimed that this is good evidence that people must have been using bows to propel them. This seems unlikely, however, as it would have involved simultaneous leaps forward in two separate technologies: adding fletching to keep the arrows flying straight; and inventing and manufacturing bows themselves. This seems a stretch. It takes around two hundred separate steps to manufacture a bow and arrow, which would probably have been beyond these people. There is actually a far simpler way that they could have propelled their arrows.

In Europe there was a long tradition of people throwing arrows using just a single piece of string, known either as Swiss arrows or

Yorkshire arrows. The user simply made a loop at one end of a piece of string and put their index finger through it. They then wrapped the other end of the string a few times around a notch at the rear of the arrow. They pulled the string taut and grasped the front of the arrow between their fingers, then simply threw the arrow with a normal over-arm sling action but releasing it a little earlier than in a normal throw. The string acted as the final element in a three-stage sling action, just like the handle of an ax. It extended the time over which people could contract their muscles and lengthened the distance over which they accelerated the arrow. The technique had the additional benefit that as the string finally unraveled from the shaft, it spun the arrow, stabilizing it in flight.

People can throw arrows vast distances in this way. The record distance for an arrow throw is said to be around 360 yards, though modern enthusiasts on YouTube and TikTok report that they can only throw their homemade arrows 110 to 165 yards (100 to 150 meters). Nevertheless, this shows that they have thrown them at 80–100 mph (35–45 meters per second) and transferred at least twenty-five to thirty-five joules of energy into the arrows, easily enough to be lethal to medium-size animals. Where I grew up in the London suburbs, none of us learned how to throw arrows in this way, but on the internet you can find many reminiscences of people my age who lived in the country and recall happy times when they used throwing arrows to hunt small mammals. They were obviously both accurate and powerful weapons.

As well as arrows, people have also long used string to throw a much wider range of projectiles. One invention that has remained in use until today is the simple slingshot, which can throw small pebbles at high speeds and over even greater distances. A sling consists of a single piece of string with a loop at one end and with a basket made of string or leather halfway along it. The slinger puts his forefinger into the loop, inserts a small stone, weighing around two ounces (60 grams) into the basket, then grips the far end of the string between his fingers.

Holding the doubled string taut with his other hand, the slinger releases the stone and whirls it around with a throwing action, before releasing the end of the string at exactly the right moment to propel the stone toward the target. The action is hard to learn, but experienced slingers can throw two-ounce (60-gram) pebbles well over two hundred yards (183 meters). Mercenary slingers, who were brought up to the task on the Balearic Islands, were an integral part of Roman armies. They used long slings that could impart one hundred joules of energy into their torpedo-shaped lead sling stones and throw them up to four hundred yards (365 meters), raining death and terror onto their enemies. Shorter slings are less powerful but more accurate than long ones and have traditionally been used by hunter-gatherers to kill small animals. Most famously, shepherds such as the boy David from the Old Testament used them to scare away wolves and kill passing giants. And the climactic scene in Khaled Hosseini's bestselling novel *The Kite Runner* shows that slings remain popular and lethal weapons with Afghan children to this day.

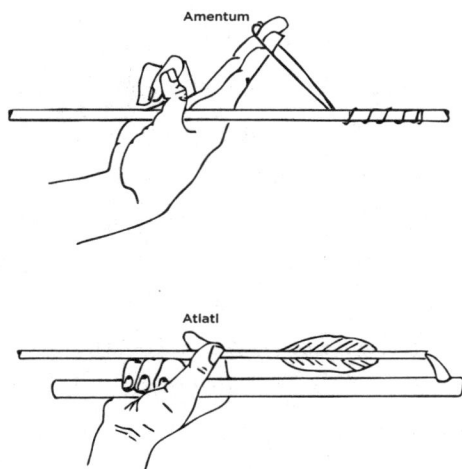

Two methods of throwing arrows or darts farther. The amentum (top) uses a string to accelerate and rotate the dart. The atlatl (below) uses a short length of wood or antler.

The sling technique was also developed in the Upper Paleolithic to throw darts that were larger and heavier than arrows, but lighter than spears, to hunt large animals. Projectiles that weigh four to five ounces (110–140 grams) are too heavy to launch with a string wrapped around a single finger. At the beginning of the twentieth century, however, French archeologists unearthed what initially appeared baffling items from Upper Paleolithic caves in the Dordogne region: small lengths of antler with a hole toward one end. They were known as *batôns de commandement*, since they appeared to have no obvious use, and archeologists assumed that they had a symbolic role rather like a scepter. Other archeologists suggested that they were spear straighteners, though they have not explained how they would have been used for this task.

By far the best suggestion has come from enthusiasts of primitive technology, who have shown that Paleolithic hunters could have tied string through the hole in the baton and held it by the other end, using it as a handle for dart throwing. These enthusiasts have shown that using this apparatus they can launch four-ounce (110-gram) darts at speeds of 60 mph (27 meters per second), giving them up to fifty joules of kinetic energy and throwing them around seventy yards (65 meters). String-propelled darts would have been formidable weapons. An engraving on a baton from the famous cave of La Madeleine, which dates from twenty thousand years ago, shows a hunter holding a dart as one would if about to throw it, grasping it toward its front end. Greek hypaspist infantrymen used lightweight javelins in classical times, propelling them using leather thongs instead of string and holding a loop at one end of the thong with their first two fingers.

String-based propulsion is clearly effective, but it has the disadvantage that the thrower has little control over the movement of the string and cannot use their wrist to add any extra power. Later in the Upper Paleolithic, therefore, people developed a new technique to throw darts using a six-to-ten-inch (15–25 centimeter) length of wood or antler—a spear thrower. The mechanics of the spear thrower are essentially the same as those of a throwing string, but it has to be used rather differently.

The user holds the spear thrower at one end and then hooks a notch at its far end over the rear of the dart. They then grasp the dart toward its rear end and throw it with an overhand throwing action. They can also flex their wrist toward the end of the action to propel the dart more accurately and powerfully in the right direction. Practiced experimental archeologists can deliver fifty joules of energy to four- to eight-ounce (110- to 220-gram) darts, the same as using *bâtons de commandement*, throwing them at around 60 mph (27 meters per second) and up to seventy yards (65 meters). Spear throwers are such effective weapons that they are still used by hunter-gatherers in Australia, where they are known as woomeras, and in Central America, where they are known as atlatls. Their main downside is that they do not spin the dart in the way that string-based propulsion systems do, so users have to fletch their darts like arrows to keep them flying straight.

Wooden spear throwers were such effective weapons that they helped lead to the demise of many large tundra- and steppe-dwelling mammals toward the end of the Upper Paleolithic. And as the ice melted, forests spread across the northern temperate regions. They would have been more difficult to hunt in and would have harbored far smaller mammals than the open steppes. Having already had to add fletching to their darts, people had to revert to using smaller fletched arrows to hunt, and to propel them more accurately they invented a totally new form of string-based power weapon—the bow. Rather than extending the time and distance over which our muscles can act, as sling-based weapons do, bows use the principle of elastic energy storage. Archers pull back on the string of their bow, bending its arms to store elastic energy, energy that is quickly released when the archer looses the string. Archers cannot supply as much energy to their arrows as a slinger can to his stones because they only contract the muscles of the arms and shoulders, not those in the trunk and legs. However, because they contract these muscles more slowly, each gram of muscle can produce more energy, so an archer can still store around thirty joules of energy in the arms of their bow, enough to accelerate their one-ounce (28-gram) arrows to speeds of up to 120 mph (55 meters per second) and propel them upward of two hundred yards (183 meters). And bows have the further advantage that the archer moves much less than a slinger, so they are far better stealth weapons.

By the Mesolithic period, people were therefore equipped with a whole armory of handheld tools and long- and short-distance projectile weapons. They had become a superpredator. And the development of the bow also kick-started a final flurry of inventions that gave Mesolithic people an almost-total mastery of the material world around them.

When I was young, children's adventure stories used to be filled with tales about castaways or explorers who lived on "roots and berries" or "hips and haws," and who started fires by "rubbing two sticks together." Unfortunately, the books never gave any details about how they actually did this, and I never met anyone who had ever started any fires in this way. It's easy to see how rubbing sticks together should work; the friction should transform mechanical energy into heat. However, the difficulty in actually starting a fire is to restrict the contact to a sufficiently small area, to move the sticks fast enough, and to apply large enough friction forces to heat the timber up to a temperature at which it will spontaneously combust. Rubbing the sticks back and forth against each other is simply not fast enough to work. It is far more practical and economical to hold one stick with its end resting on the side of the other and to rotate it. You simply hold the stick between the palms of your hands and sweep them back and forth as if you were rubbing them to keep warm. This works, but with both hands being used to hold the stick, it is impossible to push down on it at the same time, so the friction force and energy generated will be limited. One solution hunter-gatherers came up with was to use a bow to spin the stick. The fire lighter wrapped one loop of the bow's string around the stick and simply swept the bow back and forth, like a violinist playing a fiddle. Not only does this rapidly rotate the stick, but the fire lighter can also use their free hand to press down on the top of the stick with a concave stone to maximize the friction and hold the stick steady.

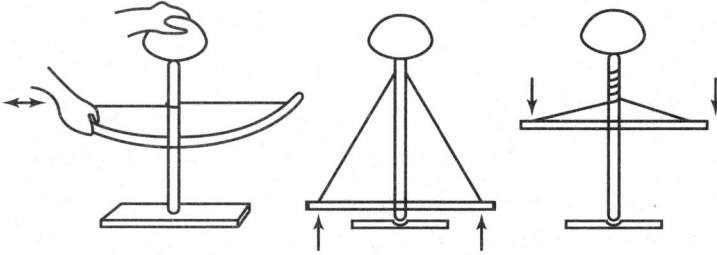

Bow (left) and pump drill used by both Neolithic and modern hunter-gatherers. In both, a single user can rotate a wooden rod to start fires or drill holes.

The idea of the fire drill was born, and people quickly learned how to make one that was even more effective: the pump drill. Hunter-gatherers can make them in just a few minutes. In the first stage of manufacture, the fire lighter drills a hole in a short length of wood, ties the two ends to a piece of string, and inserts the middle of the string into a groove at the top of a drill stick. They then rotate the stick until the string is wrapped several times around it and the plank is raised to the top. The fire lighter then pushes the plank up and down, while pushing down on the top of the stick, rotating the stick back and forth. Hunter-gatherers no longer had to carry rare lumps of iron pyrites around with them; they could light fires almost anywhere.

People also realized that they could use drills to cut perfectly circular holes not only in wood but also in much-harder materials. They simply needed to cover the tip of the drill with sand or attach a hard-wearing flint tip. They could perforate shells or beads to produce necklaces or drill into larger pieces of jet to produce attractive rings. And they could perforate much-larger pieces of stone to produce yet another power tool—a hafted hammer. They could flake a nodule of stone into the shape of a hammer head and drill a large hole all the way through its center and insert a wooden handle. Armed with such a powerful hafted hammer, people would have been even better at cracking open nuts, shells, and bones. And combined with a chisel made of stone, or a wooden one with a beaver-tooth tip, they could cut wood cleanly across the grain. It would help them construct better wood joints.

Women also took advantage of drills to make the small perforated disks of stone that archeologists call whorls. They attached them to a wooden rod to create a tool that looks just like a spinning top—the drop spindle—and used them to speed up the production of string and thread. The spinner simply joined one end of a bundle of wool, flax, or cotton to the top of the spindle and dangled it from their hand. They then gradually teased out the fibers and set the spindle spinning to twist the fibers together, before stopping the spindle and winding the length of spun thread onto it. They then had to repeat the process almost ad infinitum throughout the rest of their lives, making the yarn they would subsequently weave into cloth.

By the end of the Mesolithic period, around ten thousand years ago, people had developed the vast majority of the human-powered tools we still use today. They used a whole range of short- and long-distance weapons, from clubs, whips, spears, and bows. And they used all of our most common shaping tools: axes, adzes, hammers, chisels, knives, and drills. They had the wherewithal to exploit the natural environment wherever they were in the world. From the rainforests and savannas of the tropics, to the grasslands and broad-leaved woodlands of temperate regions, and even to the boreal forests and tundra of the Arctic, they tailored their technology to the local conditions. They could kill virtually any animal and carve it up. They could collect and process a whole range of plant foods, from nuts, fruits, and seeds, to underground roots and tubers. They could exploit a wide range of materials, from wood, stone, and antler, to bone, leather, fiber, and ceramics. They could fell small trees and use their timber not only to produce their weapons and tools, but to build houses. They could hollow out logs to make canoes and construct larger boats by covering a wooden framework with animal skins. They could light fires to keep themselves warm, cook their food, and make their glue.

With these impressive skills, it is not surprising that by ten thousand years ago modern humans had colonized all the continents bar Antarctica, found ways to thrive in virtually every habitat, and increased their population to an impressive 8 million worldwide. They had come a long way from being feeble apes. Their technological world would have been surprisingly like our own, and they had even found the key to the design of the rotating devices that were to lead to the development of wheels and to virtually all of the technology that we rely on today. And they did it all when living in small self-sufficient nomadic groups. Never again would people's wit and physicality be so well integrated. Never again would people be so fully alive or individuals be so self-sufficient and at home in the natural world.

But humans had not expanded the size of their brains for over sixty thousand years. It seemed unlikely that over the next ten thousand years they would develop an engineering civilization that would be capable of supporting a world population of 8 billion: that we would be able to feed so many people; exert such huge forces and develop such prodigious amounts of energy; produce so many goods and live so plentifully; travel so far and fast; consume so much; and rain such destruction on our planet. In the second half of the book we will see that this most unlikely outcome has resulted from a complex set of circumstances, each step in our progress being contingent in unpredictable ways on the impact of previous steps. And we will see why the process culminated on one cold, damp island off the northwest coast of Europe in a revolution in engineering that would produce an industrialized, urban country, spread across the world, and exalt people's standard of living to undreamed of heights, before finally threatening our very existence.

Part Two

EXPANDING OUR POWER

Chapter 8

GROWING OUR POWER

I'll always remember the shock and disbelief I felt back in 1990 when I first beheld the vast expanse of North America's grain belt. I was flying to the small city of Regina, Saskatchewan, for an interview for an academic post but was overwhelmed by what I saw as our plane flew in from Chicago. Dead-flat checkerboard fields stretched as far as the eye could see, uninterrupted by any hill, hedge, or tree. The cereal fields of the Great Plains stretch three thousand miles from the wheat fields of the north to the cornfields of the south and cover around 1.1 million square miles (2.9 million square kilometers). They make the wheat fields of England's notoriously flat Norfolk fens look wild and mountainous. It suddenly made me realize how we humans can completely bend the world to our will and that by and large we have chosen to replace native vegetation with cereals. In Europe and the Americas, wheat dominates in cool temperate areas, with barley and oats becoming more common in the cold north, rye in areas with a dry continental climate, and corn or maize in the hotter South. In Asia, farmers grow wheat in the north, and rice in the warmer, wetter subtropics, while in Africa farmers tend to grow millet, sorghum, or teff.

All around the world, cereal farming is big business. And cereal farmers seem to be much wealthier than cattle or sheep farmers, as my first PhD student Mitch Crook and I found when we attended the

Cereals '94 trade event to publicize our research on how to prevent wheat from falling over, or lodging. The previous autumn the organizers had planted a massive area of Cambridgeshire with plots of cereals, to showcase different crop varieties, while in between the plots there was room for ranks of gigantic tractors, sprayers, and combine harvesters and marquees filled with the trade stands of seed and agrochemical companies. Range Rovers and Land Cruisers filled the car park, and the exhibitors were giving away plenty of merchandise. One company's stand was serving free hamburgers, something that would be unheard of at a scientific conference, and that was therefore irresistible. I felt a bit greedy going up for a second hamburger; being younger, hungrier, and more brazen, Mitch had five. Seeing all the money on display, I was irresistibly reminded of John Crisp's 1980s comic song "I've Never Seen a Farmer on a Bike."

So it is all too easy to believe the story that has long been told about the birth of the modern world: that the advent of cereal farming was the key event that raised human beings' productive capability and led inexorably to the rise of civilization. But other observations tell a totally different tale. Cereals are so hard to cultivate and unpalatable, and they produce so little grain, that gardeners and allotment holders alike never bother to grow them. They concentrate instead on fruit and vegetables, which they can grow in much greater quantities; which are far less trouble to cultivate; which are easier to cook; and which are much nicer to eat. Indeed, recent studies, summarized by Dave Goulson of the University of Sussex, have shown that even an unskilled allotment holder can produce four times as much food on a given area of land as a cereal farmer. This most basic observation immediately casts doubt on the story we have always been told about the superiority of cereals; and when we examine the archeological evidence, it blows out of the water the simple story that cereal farming led inexorably to the rise of civilization.

For a start, Mesolithic hunter-gatherers already had all the tools and skills that they needed to produce what are generally accepted as the hallmarks of civilization. They were able to create beautiful artworks, play music, and produce ritual objects that showed they practiced religions. In the eleven-thousand-year-old settlement of Göbekli Tepe in Anatolia, modern-day Turkey, they had even built a city, with stone dwellings, statues, and shrines. Other late-Mesolithic structures such as the megaliths and stone alignments at Carnac, Brittany, show that the people had superb organizational and engineering skills and were capable of moving stones weighing up to forty tons. They were already civilized, having both supreme artistry and superb engineering skills.

The hunter-gatherer lifestyle had drawbacks, however. Because hunter-gatherers use up the resources around where they are living, they have to continually move about in their search for food. They cannot settle down easily in one place, build a permanent house, or amass large numbers of tools and other possessions to make their lives easier. Their lifestyle inevitably involves a great deal of movement. The men of the Hadza tribe, for example, walk an average of eight miles a day on their hunting expeditions, while the women travel around five miles in their search for roots and berries. Consequently wherever there is a sufficiently concentrated source of food, people have always preferred to settle down in permanent villages. In the Pacific northwest of America, for instance, the Chinook live in fishing villages dominated by huge timber longhouses and totem poles. And ten thousand years ago in Yorkshire, England, the Mesolithic hunters were able to set up a semipermanent village around a lake at Star Carr, where they could fish and hunt deer.

In some of the warmer, wetter regions of the world, hunter-gatherers also settled down inland where the excellent growing conditions enabled them to cultivate high-yielding fruits and root crops. A modern example, the Tsimane people of Bolivia, produce most of their food in gardens of manioc, bananas, and wild rice. They cultivate the ground with mattocks—tools rather like upsized adzes—and supplement their vegetarian diet with a little part-time hunting. Research led by Herman Pontzer of Duke University and his colleagues has shown that the Tsimane spend only four hours per day on subsistence activities, around 30 percent less than the Hadza, yet they obtain on average 15 percent more energy. Pontzer's group also examined the research that other anthropologists have carried out on the time and energy budgets of hunter-gatherers and horticulturalists from all around the world. These studies show that the Tsimane and Hadza are far from unique. Gardening is almost always an easier way of making a living than hunting and gathering.

Small-scale horticultural societies are common around the edges of tropical forests in the Amazon, West Africa, and Southeast Asia, where people can cultivate fruit trees such as bananas, plantains, and palms; dig up the swollen roots of herbaceous plants such as yams, sweet potatoes, and manioc; harvest vegetables such as tomatoes, peppers, and squashes; and make use of giant grasses such as corn and sugarcane. To develop a garden all they need is to clear a small area of ground around their settlement and burn the existing vegetation to incorporate its fertility. The land is then ready for them to dig holes using digging sticks or mattocks and to plant the seeds, cuttings, or tubers of the plants they want to grow. They cultivate a mosaic of different plants around their plot and so can harvest food year-round and keep a constant cover of vegetation, maximizing yields and minimizing soil erosion. Horticulturalists' gardens strongly resemble the plots of modern-day exponents of permaculture. The only downside to this form of swidden or slash-and-burn cultivation is that the soil's fertility quickly drops, so after a few years, the gardeners have to cultivate a new plot and give the old one time to recover. This sets a limit on the number of people an area of land can support.

There is good evidence, however, that in the Amazon horticultural societies overcame even this drawback by maintaining soil fertility. In recent years, archeologists have been finding huge numbers of mounds where the soil had been built up with charcoal, compost, and manure to create areas of so-called Amazonian dark earths, which the people could have cropped continuously. Recent satellite observations using lidar technology, which can map the surface of the ground beneath trees, has spotted well over ten thousand such settlements. Many of these mounds were arranged together to form large-scale towns or even garden cities, which had grand central buildings, roads, and parks and covered many square miles. One of the largest was found in Ecuador's Upano Valley by Stéphen Rostain, of the French National Centre for Scientific Research (CNRS). This two-thousand-year-old garden city had more than six thousand earthen platforms arranged in a geometric pattern and connected by public squares and roads that linked fifteen separate urban centers. The population of this settlement may have numbered tens of thousands.

The evidence is also mounting that some of these towns may have been founded as long as ten thousand years ago, and that the whole Amazon basin was once densely populated. Certainly, the first European explorers in the area were amazed by the magnificence of the settlements that they discovered. The Spanish friar Gaspar de Carvajal, for instance, who accompanied an expedition from the Andes down the Amazon in 1541, reveled in the teeming throngs of Indians and noted the fear that his companions felt at beholding the grand buildings. Sadly, however, we know little about these settlements because expeditions such as Carvajal's unwittingly brought death to the indigenous people from European diseases such as smallpox, influenza, measles, and syphilis. These quickly decimated the inhabitants, who had no natural immunity to the infections, so that the cities were rapidly abandoned. They were swiftly swallowed up again by trees so that they soon reverted to jungle, which we have until recently always assumed was a primary rainforest.

The only large-scale horticultural societies in America that we know much about, therefore, are those that developed farther west and north: the Incas in the Andes and the Aztecs of Mexico. And there is little detail even about these societies since they were rapidly destroyed by the Spanish conquistadores, who vanquished them by battle and disease in the early sixteenth century. These peoples were the last of a series of empires whose inhabitants grew fruit and a wide range of vegetables: potatoes, tomatoes, sweet potatoes, manioc, squashes, peppers, and the giant grass corn. Like the modern-day Tsimane they grew their plants in small fields or garden plots, using mattocks or hoes to cultivate the soil. And despite what we regard as the primitiveness of their technology—they lacked what Europeans think of as basics such as plows, metal tools, draft animals, wheels, and ships—they built huge cities and ruled over large empires.

The productivity of their horticulture was so high—modern-day farmers and allotment holders growing the same sorts of vegetables in much the same way can produce over sixteen tons of vegetables per acre (40 tons per hectare) every year—and the effort needed to grow them so low, that a small number of gardeners, working even a small patch of land, could feed a large population. Consequently, the population density of these regions was higher than in Asia and Europe, even after thousands of years perfecting cereal cultivation. Before their demise in the early sixteenth century, the Aztecs numbered some 5–6 million over the eighty thousand square miles of their mountainous empire, giving a population density of seventy-five per square mile (29 per square kilometer). This was higher than the fifty to sixty per square mile (19–23 per square kilometer) who lived at the time in England, a rich, lowland country that in 1500 was one of the most densely populated countries in Europe. The huge Aztec island capital of Tenochtitlán, whose population rose to over two hundred thousand at its peak, was fed by a dense intercropped area of wetland that covered just three thousand square miles. The wetland was irrigated and drained by a massive series of canals, aqueducts, and ditches that enabled the gardeners to keep a constant crop cover and produce multiple harvests every year. This area supported 1.5 million people, giving a staggering population density of five hundred people per square mile (188 per square kilometer).

Using these intensive horticultural techniques, a relatively small number of farmers could support a laboring workforce that was large enough to construct the magnificent stone palaces, temples, and pyramids of their cities; an army that could protect the society, conquer their neighbors, and provide a constant stream of victims to sacrifice to their gods; craftsmen to create beautiful ceramics and jewelry; a bureaucracy that could organize trade; a priesthood who could conduct their barbaric rituals; and an aristocracy who could rule over the rest. The Aztecs even developed their own forms of writing to record events. In other words, these societies had all the trappings of what we call civilization. Their lack of sophisticated machinery was little problem because of the size of the workforce they had available. The only vulnerability of the Central American civilizations seems to have been the large-scale droughts that destroyed the Olmec and Mayan empires long before Columbus's arrival in the New World. If Europeans had not come on the scene five hundred years ago, the Aztecs and Incas might still be in power and be living in much the same way as ever. The main limitation to the New World empires was their inadequate transport. Because they had only pack llamas, dugout canoes, and reed rafts to transport goods, the size of their empires was limited, and the lack of ships kept them landlocked. Nevertheless, they still managed to build impressive pyramids by transplanting large stones with sleds and hammering them to shape with jadeite rocks.

Away from the wet tropics it was not so easy to cultivate gardens. In the great northern deciduous and boreal forests of Europe, Asia, and North America, there were few plants that people could have grown. The few species of fruit trees only produced food in the autumn, so the people would have starved over the rest of the year. Consequently they retained their hunter-gatherer lifestyle, surviving principally by hunting game. In the treeless savannas, prairies, steppes, veld, and tundra that covered warmer, drier areas of the globe, there were few plants that were edible to human beings at all: just perennial grasses and lichen. Here, people had to follow the herds of grazing ungulates that roamed about them: caribou, reindeer, cattle, or horses. They became herders, learning how to exploit their animals not only for food but also for milk, blood, and, in the case of reindeer and horses, for transport. They carried their homes and belongings along with them, dragging them behind them on the first sleds, travois, which they made using the crossed poles of their tents. And they learned to control, catch, and tame their animals, using simple sling tools such as sticks, whips, lassos, and bolas. Like the technology of horticultural societies, that of the herders has changed little since the end of the last ice age, and if our civilization had not expanded so rapidly in recent years and endangered their surroundings, they would still be thriving today.

Strange as it may seem to us, perhaps the most difficult places, apart from deserts, for people to survive in were the areas where cereals were first farmed and which historians have long described as the birthplaces of civilization: western Asia and northern China. The mountainous land around the eastern Mediterranean, for instance, the Levant and Anatolia, modern-day Israel, Syria, and Turkey, were drought ridden. Summers were hot and dry, and plants could grow only during the wetter winter months. The sparse vegetation was dominated in the hills by inedible evergreen shrubs and in the valleys by small annual plants such as emmer wheat, barley, goat grass, and various wild legumes. Hunter-gatherers survived in the region by stalking gazelle and gathering seeds from the grasses and legumes when they ripened in late spring. But as the climate became more arid around ten thousand years ago, even this strategy became unviable, and the people turned to what we know today as cereal farming. They settled down, cultivated the soil, grew the annual plants as crops, and harvested and stored their seeds as grain.

Their main problem was that, contrary to what most historians and agronomists have asserted, cereals are pretty poor crop plants. True, the seeds have quite a high protein content, around 12–14 percent and they are easy to store. But cereal grasses produce only small quantities of grain; they are difficult to harvest, process, and eat; and they are extremely hard work to grow. I certainly found this last aspect a shock in the early 1990s when I was a young academic in my first permanent job and was carrying out research on how to prevent wheat from falling over, or lodging. In obtaining my PhD, I had acted rather like a hunter, catching flies with a butterfly net so I could film them in flight and examine their wing structure. Though this required rapid identification skills, fast reflexes, and a high embarrassment threshold, it was hardly strenuous; to find the flies I merely had to stroll around the University of Exeter campus or the nearby countryside. For my postdoctoral research fellowship on root anchorage, I acted like a gatherer or horticulturalist. I went on car trips, driven around North Yorkshire by my colleague Alastair Fitter, digging up the annual weeds that thrive in the disturbed soils of roadworks, to investigate the shape and size of their root systems. I also grew a wider range of annual weeds in pots and cultivated sunflowers and maize in the University of York's walled garden, so I could uproot them and investigate their anchorage mechanics. Obtaining my experimental subjects had been easy.

But growing wheat was a very different matter. Getting the wheat in the ground required a complex series of operations and heavy farm equipment. Consequently we—I use the academic *we*, which really means *they*—had a lot of work to do. I was fortunate that *we* in this case was my assistant, Mitch, who already had experience in setting up field trials, and who proved adept at enlisting the help of Bernard, our groundsman, and John, the neighboring farmer. I only came along to our botanical grounds at Jodrell Bank, Cheshire, after the soil had been cultivated, to help plant the field trials. I sat behind Mitch on the tractor operating our precision seed drill.

Cereals are such poor crop plants because of their ecology. Like other winter annuals, they have evolved to survive in regions with a short growing season and so are small, short-lived plants. They take only four months or so to grow, flower, and set seed; but each cereal plant produces just three or four thin stems or tillers; and their seed heads are small, making up less than 30 percent of the biomass of the plant. As a result the yields of cereals are far lower than those of root crops or fruit trees, which have longer growing seasons, and which divert almost all their energy at the end of it into their fruits or storage organs. Even today the highest-yielding wheat produces just three tons per acre (8 tons per hectare), a fraction of the yield produced by a field of potatoes or turnips. The earliest wild grasses would have yielded less than one ton per acre (2.5 tons per hectare) at most. And before grain is processed, the seeds are virtually inedible. They are coated with a protective husk and bear a long indigestible spine, or awn, which helps the seeds disperse. As anyone who has walked along a field of wheat or barley will have found, the grain presents nowhere near as palatable a snack as a banana or an apple; cereal farmers never fear being raided by children, as the owners of orchards do.

The harvesting of grain is also long and arduous, largely because of a mutation in domesticated cereal crops that most authors have extolled as one of their chief virtues. Unlike wild cereals, the ears of cultivated varieties do not shatter; over the earliest days of cereal cultivation, people inadvertently selected for plants bearing seeds that stayed attached to the ear, even when they were ripe; they were less likely to collect seeds that fell off earlier, so plants with shatter-resistant seed heads soon become the norm. Growing cereals with shatter-resistant ears ensures that the farmer does not lose any grain before it is harvested, but it also makes harvesting the crop difficult. Early farmers had to invent a whole new range of tools and develop novel techniques to remove the grains from the rest of the plant and separate the wheat from the chaff. To harvest a cereal they could not just pull the seeds from the stem as you would harvest an apple; they had to cut through the stems with a special curved knife—a sickle—sweeping it sideways to slice through the bottom of the stems while they gripped the tops to hold them taut. Reaping is a graceful, rhythmic action, but it is energy- and time-consuming, and because the reaper had to stoop down to the ground to cut through the bottom of the stems, it was also backbreaking. The reaper also had to be followed by a binder, who gathered up the harvested plants and tied them together into a sheaf.

But reaping and binding was merely the start of the hard work. To remove the grains from the stems of a cereal it has to be threshed, a process that is even more laborious and time-consuming. Threshing rice is not too difficult. The thresher holds a sheaf of rice at the base and wields it like a club, using a sling action to slam the ears of the rice against a rock or a plank of wood. For wheat and barley, the threshing is more complex. To hand thresh wheat, farm laborers had to spread the harvested plants across the dry floor of a barn and hit them repeatedly with a flail—a simple wooden device that consists of a yard-long (ninety-centimeter) hand staff tied to a two-foot-long (60-centimeter) beater. The thresher powered the flail using a four-stage sling action—rotating the shoulders in turn powered the swing of the upper arm, forearm, hand staff, and beater—giving the tip of the flail enough energy to break off the grains, something in the region of one hundred to two hundred joules.

Threshing is arduous, and since our ability to supply oxygen to our muscles limits our sustainable power output to just eighty watts, an individual can at best thresh 45 to 65 pounds (20 to 30 kilograms) of cereal per hour. Until the advent of modern machinery, threshing took around a quarter of all the labor time of the peasants and occupied them for the first half of the winter. Even when the grain was threshed and the wind had blown away the chaff, a process known as winnowing, the raw grain was still largely inedible—as anyone who has tried to eat grain plucked from a field of wheat will know. Grains are solid lumps of dry starch and have to be broken up into smaller pieces or ground into fine flour and then baked or boiled in water to make them palatable. And milling grain to produce flour demanded yet another invention—the quern.

The simplest way of breaking up cereal grains is to place them in a stone pot, or mortar, and hit them repeatedly with the end of a wooden stick, or pestle. Unfortunately, this also uses a great deal of energy. A rather better way is to grind it using a milling tool or quern. The simplest form, the saddle quern, first appeared in the Middle East around twelve thousand years ago. It consists of two pieces of stone: a large flattened bed stone, on which the miller strews the grain; and a smaller, elongated stone or rubber that the miller holds at either end and rubs back and forth over the bed stone to crush the grains. It sounds easy, but milling grain in this way would have been slow. A skilled miller would have been able to produce only around two pounds (1 kilogram) of flour per hour, and it would have been an extremely arduous operation. The millers—at this period exclusively women—had to kneel down before the quern and push the rubber back and forth, using the whole body as a huge crank to provide the power. And since much of the energy the women produced went into moving their body back and forth, rather than just squashing the grain, milling, like threshing, was another energy-intensive task. Moreover, because of the unnatural body position they had to adopt, Neolithic women were prone to arthritis, particularly of the big-toe joint. Even after milling the grain, the job was not complete. The earliest cereal farmers had to mix the flour with water and other ingredients and heat it over a wood fire to break down some of the starch, to produce a form of bread.

If harvesting and preparing cereals demanded ingenuity, time, and energy, the greatest difficulty in farming cereals lay in actually growing the plants. Unlike fruit trees or root crops, you cannot plant individual cereal seeds and cultivate each plant separately; the plants are just too small. To make it worth your while, you have to cultivate thousands of plants together to produce a monocultural block or field. And before you can plant your field, you have to till the whole area so that the tiny seeds can germinate; you have to open up the whole soil surface and break apart the clods to form a fine seedbed. Digging sticks and mattocks, which are perfectly good for cultivating a single large plant, are ill-suited to this task. After all, it takes a lot of energy to use a mattock. You not only have to accelerate it on its power stroke, generating up to a hundred joules of kinetic energy, but you also have to lift it backward and upward to prepare for the next stroke, all the while leaning forward in an unnatural posture. Since each stroke of the mattock opens up just a small area of soil, tilling is extremely time-consuming. In an experiment carried out in 2023, for instance, it took two hundred man-hours for members of the Prehistoric Workshop of Leicestershire to prepare just one acre (0.4 hectares) of soil using this technique. The problem is all the more pressing because cereals have such low yields. Early cereal farmers, who had to cultivate far larger plots than the horticulturalists of the tropics, would have had to spend far more time and effort cultivating the ground; one day's labor would only be enough to grow 130 pounds (60 kilograms) of emmer wheat. Early farmers would have had to spend weeks preparing the soil each year just to feed themselves.

There is one final difficulty with cereal cultivation: weeds. Being small plants, cereals are easily swamped by annual weeds that can grow up between them and shade them out. And because early farmers planted the crop by scattering their cereal seeds by hand, their crops would have been extremely hard to weed without damaging the cereal plants themselves. The farmers probably used long-handled hoes to cut the weeds off at the base, like medieval peasants or modern gardeners, but the task would have kept them busy throughout the growing season.

So the conventional story that historians tell about how the benefits of cereals gave rise to civilization is highly implausible. The first cereal farmers would have had to work far harder and for lower rewards than horticulturalists, or even hunter-gatherers. The evidence from their graves tells us that they were significantly shorter and suffered from more disease than their counterparts elsewhere. Consequently for hundreds of years cereal farming was limited to the heartlands where it first emerged. In the Levant and the hills of Anatolia, farmers lived in small villages that numbered in the hundreds rather than thousands. There seemed to be no reason why their way of life would ever become the dominant one on the planet.

On the other hand, neither was early cereal farming a disastrous poverty trap. Cereal farmers may have had to work harder than most other humans, but they survived in arid areas where hunter-gatherers would have perished. Excavations at early farming settlements, such as Catalhöyük, Anatolia, which was founded around 7500 BCE, show that the people lived in near-identical houses and worked together tending their small plots just as modern-day horticulturalists tend their gardens and orchards; their society was just as egalitarian. It was actually the arduousness of their working lives that stimulated cereal farmers to find new ways to reduce their workload and increase their productivity. It set them on a course of engineering innovation that would paradoxically lead cereal farmers to overcome the disadvantages of their crop plants, become wealthier, expand the area they farmed, and eventually dominate the entire world.

Chapter 9

REDUCING THE WORKLOAD

If you want to get a feel for people's way of life toward the end of the Neolithic period, when they were still farming cereals using stone tools, there is no better place to visit than the village of Skara Brae. A collection of circular houses looking out over the western shore of Orkney's Mainland island, it owes its miraculous survival to its having been built on this almost treeless island out of the local sandstone rather than the more usual wood, and having been buried by a sandstorm around 2600 BCE, preserving most of it intact. The remains of the houses, and a recent reconstruction of one of them, show that the dwellings would have been remarkably snug and well-appointed, with their thick earth-packed walls, a central hearth, bed boxes around the walls, inglenooks for storage, and even a latrine. Each house was also furnished with a stone dresser opposite the door, and a range of carved stone ornaments. Fitted with a wood-framed roof, central fire, and animal-skin hangings, the reconstructed house gives off a strong sense of that most Danish quality, hygge.

One gets the feeling that these people, who farmed barley and herded cattle and sheep, would have led comfortable lives, despite their exposed location on the very edge of the Arctic Ocean. Their skeletons show that they were healthy, being as tall as modern people, and other finds show that they wore leather clothes festooned with jewelry;

fashioned tools from stone and bone; fired ceramic cooking vessels; and played with finely crafted gaming pieces. They made many if not most of their items in their on-site workshop. And they were not as remote from the center of events as you might suppose. They were living just a few miles from some of the most spectacular Neolithic ritual sites in Europe: the Stones of Stenness, the Ring of Brodgar, and the Maeshowe chambered tomb. Neolithic Orkney seems to have been an important place, and as densely inhabited as it is today, a far cry from when it had been the home of the few hunter-gatherer families who had scraped a living on the islands a few thousand years before.

At first glance, therefore, Skara Brae seems to be a shining example of the benefits of cereal farming. The lifestyle enabled a large population to thrive in one of the most extreme environments of Europe. However, it is important to point out that the village was founded over five thousand years after the first cereal farmers planted their crops in the Middle East. Cereal farming had not been an overnight success, and the Mesolithic inhabitants of Europe had preferred to carry on being hunter-gatherers for thousands of years. Not until Neolithic farmers developed three new technologies—the plow, ceramic pottery, and the ground-stone ax—and completed what archeologists call the Neolithic package did cereal farming become an attractive-enough proposition to spread away from its heartlands. As we shall see, each of these technologies helped to reduce the farmers' workload, enabling them to clear and cultivate more land and improve their ability to process their food. At last they could grow more food and have larger, tastier meals. They could live better lives. But as we shall see, these new technologies also had other effects on Neolithic society that would influence the ways in which we live and think to this day.

Probably the most important of the innovations, and certainly the one that is most closely associated with cereal farming, is the ox-drawn plow. Like the herders of the steppes, the cereal farmers were quick to domesticate large herbivores. Ruminants proved to be the most useful animals since they could survive by eating plant material that was inedible to humans. The farmers could herd sheep and goats in the rough evergreen scrub that covered the hills around their fields and use them to provide meat, milk, and wool. And they could keep and domesticate the local species of bovid: aurochs *Bos primigenius* in the Near East and Europe and water buffalo *Babulus babulis* or gaur *Bos gaurus* in Asia and the Indian subcontinent. Farmers could feed the animals hay from local grasslands or straw from the cereals that they were growing, milk the cows, and slaughter the bullocks. There would be less food waste, and the manure that the animals produced would help the farmers maintain soil fertility. And cattle, being the largest of the beasts, could also be used to take over from human labor—the first example of people outsourcing their mechanical power.

We saw in the last chapter how difficult it was for the first farmers to cultivate enough land to support themselves using mattocks. They would have used far less energy if they could instead have pulled a digging stick sideways across the soil surface. Unfortunately, this would have been beyond their capability; a recent study by Irish and German experimental archeologists has shown that it requires forces in the region of 60–150 pounds (270–670 newtons) to till the soil in this way, the equivalent of a person walking continuously up 30 to 60 degree slopes. A human-powered device would have had to be dragged by several people working flat out.

A much-better solution was to yoke a cow to pull a simple framework of wood equipped with a stone tip—a plow. The farmer merely had to hold on to the rear of the plow and direct its passage through the soil, controlling the depth of the plowshare. The animal would have had no difficulty providing enough force or power. To a 1,300 pound (600-kilogram) cow or a 650 pound (300-kilogram) water buffalo, pulling a force of 60 pounds (270 newtons) is easy, as it is equivalent to their walking up a slope of just 2.5 to 5 degrees. Most draft animals can easily pull 10–15 percent of their body weight. Over a working day, a cow, walking at speeds of around 2 mph (1 meter per second), can produce an average power output of around three hundred watts, four times that of a human. Consequently a single cow could plow over ten times the area as a man equipped with a mattock, around one acre per day. This multiplied the area a farmer could cultivate and so the amount of food they could grow.

The second innovation—ceramics—helped the farmers to reduce the time and effort to cook their food. The obvious way to heat food and make it more digestible is over an open fire, and I have many happy memories of cooking basic starchy foods in just this way when I was a Cub Scout. We made a paste of flour and water, with a pinch of salt, stuck it on the end of a stick, and held it in the flames, finishing the meal with marshmallows cooked in just the same way. And we baked potatoes by coating them in foil and burying them in the ashes of our fire, a method that Mexicans still use today.

Open fires use a lot of wood, however, since the heat can escape all too easily to the environment. Excavations at the early Neolithic village of Catalhöyük show that these early farmers had found a way to bake a form of leavened bread and to do it more economically. Each house had its own dome-shaped clay oven into which they placed the smoldering embers from an outside fire. The embers produced a near-smokeless flame, and the farmers placed their dough onto heated clay tablets and slid them inside the oven, like modern-day pita bread makers or pizza chefs. But man cannot live on bread alone. Neolithic people also needed a method to cook their other foods: the pulses, peas, onions, and meat that could vary their diet and improve their nutrition. Their solution was to use pottery.

Paleolithic and Mesolithic peoples had been making ceramics for thousands of years, heating clay figurines, such as the thirty-thousand-year-old Venus of Dolní Věstonice, in fires to harden them. But open fires rarely get to temperatures above 750°F (400°C). They are not hot enough to fire pottery that is resilient and waterproof enough to act as a cooking pot or storage vessel. To produce earthenware you have to heat the clay to temperatures of over 1,100°F (600°C), which drives off all the water and forms permanent bonds between the clay particles; at even higher temperatures, the clay starts to become more vitrified and waterproof. The earliest Neolithic potters, both in Eurasia and the Americas, attempted to raise the temperature by lighting their fires in a pit. This certainly lowered heat loss, but it also reduced the airflow to the fire, so though it economized on wood, it failed to significantly raise the temperature of the fire.

The innovation that enabled potters to provide both good insulation and good airflow to their fires was the kiln, which was first devised in Anatolia around 6000 BCE. Excavations at the settlement of Yamim Tepe show that by 5300 BCE potters had perfected the beehive kiln, which continues to be used to this day. The main structure was a ceramic or stone dome that was split into two chambers. The upper chamber tapered to a small smoke hole at the top and was linked to the lower by a perforated shelf. The lower chamber was open on one side so that the potters could make a fire in it. Later kilns were more asymmetrical with a separate firebox to the side of the lower chamber. This arrangement ensured that the pots in the upper chamber were kept well away from any flames, while the powerful draw from the small hole at the top of the upper chamber pulled air past the fire and into the kiln and drove hot smoke up past the pots. With such an efficient arrangement, potters could fire their ceramics at a higher temperature while using less wood. They could make better, cheaper pots, which people could place over their fires, mixing their flour with seeds, vegetables, herbs, meat, and water to cook a wider range of broths, stews, porridges, and gruels.

Together, these two new technologies enabled cereal farmers to increase their productivity and improve their diet, but it also magnified their thirst for land. It spurred them to expand their range, away from the hills of their heartlands down to more fertile river valleys, most notably the Fertile Crescent, and the Nile and Indus Valleys. Here they used their cattle to clear the marshland of its lush reed vegetation. They started to control the water levels of these valleys with drainage ditches and canals, enabling them to farm the rich alluvial soils. Irrigation also enabled them to maintain a permanent water supply to their plants so the farmers were no longer restricted to the short winter growing season; they could grow more than one crop each year. With more grain produced, settlements expanded, and in these fertile valleys cereal farmers built the first cities. The earliest, Eridu, founded around 5500 BCE in the lower Euphrates Valley, was surrounded by the typical crisscross of irrigation ditches from which the people also obtained their chief building material, sun-dried bricks.

For the first time, cereal farming also spread into Europe, across the Balkans, and westward along the shores of the Mediterranean Sea, through Greece and Italy. The pioneers who moved into the new areas were fortunate that the Mediterranean climate, with its mild, wet winters and dry, hot summers, was ideal for the winter annuals they were growing as crops. Their form of farming could support a denser population than hunter-gathering in the region, and incoming farmers quickly displaced the native population. They also found it surprisingly easy to remove the other main barrier that might have blocked their spread: the evergreen forests that coated the region. They found that they could simply set fire to the flammable trees and shrubs, since their leaves and branches were filled with volatile defense chemicals and resins. Forest fires could also clear and prepare the ground for cultivation. By the middle of the sixth millennium BCE, farming had reached southern Spain.

It proved harder, however, for cereal farmers to colonize northern Europe. The first problem was that it took time for their crops to adapt to the cooler, wetter climate, with the growing season limited to the summer rather than winter months. The spread was also blocked by a more formidable foe: the deciduous forests, which, unlike those farther south, are too damp to burn. The farmers had to develop a third technology to remove them and make what has become the poster boy of the Neolithic in Europe, the ground-stone ax.

At first glance ground-stone axes don't appear to be a great advance on the flaked tranchet axes of the Mesolithic. They certainly look more beautiful, with their smooth, shiny surface, but their blades are blunter, and they are broader. They don't look as if they could cut through wood at all. And the heads were made from quite rare types of fine-grained stone such as chert, jade, or sandstone, so that they would have had to be traded over long distances. In Britain, over a quarter of all ax heads were made from the greenstone that was mined from the fells of Langdale in the Lake District. In what is known as the Langdale ax industry, miners quarried the blanks and roughed them out using the traditional flaking technique. They were then polished nearby using local sandstone blocks as whetstones and exported around Britain and Ireland. Grinding alone takes ten to twenty man-hours, so at first glance it hardly seems worthwhile.

However, when used to fell a tree, ground-stone axes are far superior to tranchet axes. Because of their smooth surface, they are less likely to break and seldom get stuck in the wood. Experimental archeologists have shown that a group of twenty people using ground-stone axes can clear an acre of oak woodland in anything between one and four days, depending on the size of the trees. Experiments that I carried out with my project student Joao Oliveira showed why they are so effective. The broad, smooth blade works just like a modern splitting maul; it levers the wood cells apart along the grain and slides through the wood surprisingly easily.

Being able to clear the forest finally gave Neolithic farmers the ability to spread across northern Europe and Asia. From 5400 BCE for instance, the Linearbandkeramic (LBK) people swept from the Balkans to the Atlantic coast in just a few centuries, taking with them their characteristic banded-patterned ceramic pots after which they are named. Their ground-stone axes and adzes also proved to be excellent woodworking tools, and the LBK people became skilled carpenters. They hewed felled tree trunks into square beams that they could join together to build the frames of their huge longhouses, and they split beams along their length to produce planks that they could lay together to form the walls. And after they had felled a tree they could harvest the long, straight shoots of wood—coppice poles—that sprouted from the stump and weave them together to manufacture doors and internal partitions and build fences and paddocks to house their animals. They could use the material that was growing all around them to create wooden villages in regions where ones made out of mud bricks would simply have washed away.

Ground-stone tools spread rapidly around Eurasia, and since the stone they were made from was only found in a few locations, the ground-ax industry became part of a new long-distance market economy. Archeologists have found large numbers of ax heads all along the banks of Europe's major rivers, the Rhine and the Danube, for instance. This indicates that the axes, along with other goods, were transported by water across the continent; since the headwaters of the two rivers almost meet in the Black Forest of southern Germany, they link up the Atlantic coast of Holland with the Black Sea coast of Romania and Bulgaria. The ground-stone tools were also crucial to the new trading networks in another way because they helped shipwrights build bigger boats than ever before. The shipwrights could hollow out huge trees to build log boats for river transport and weave together coppice poles into the framework for even-larger skin-covered boats, which enabled traders to carry goods across the sea. It was the beginning of the rise of globalization.

By 4000 BCE the engineering advances that made up the Neolithic package had overcome at least some of the disadvantages of cereal farming. Cereal farmers had spread across much of Europe and Asia, bringing their expertise in ceramics and woodworking with them, along with their draft animals and cultivation tools. By cultivating the land, farmers could support a far greater number of people than the hunter-gatherers who they displaced. Meanwhile, in the warmer, more fertile river valleys of the Middle East and Asia, farmers were continuing to improve their irrigation techniques. They had still not reached the level of productivity that the Central American gardening civilizations would later achieve, so their societies were as yet less urbanized. However, they were far more linked together by long-distance trading networks and had a higher level of engineering skills.

But the cereal farmers' unique dependence on engineering and technology also set Eurasia on a new social and political course. Since plows allow farmers to cultivate bigger plots, farming started to cover the entire area around villages and towns, cultivated fields replacing the natural vegetation. This made land rather than labor the limiting factor. Land became more valuable and labor less so, and this decoupled wealth from labor. The economist Samuel Bowles of the Santa Fe Institute has shown that by the fifth millennium BCE some farmers were wealthy enough to raise and keep specialist plow oxen (castrated bulls) rather than general-purpose cows. Consequently they could cultivate more land and hire out their beasts to poorer farmers and become even wealthier. Since wealth could be passed on, social inequality rose rapidly and with it the emergence of a wealthy landowning elite who were elevated above the craftsmen and small farmers. And soon another technological revolution—in chemistry and materials—would enable Eurasian craftsmen to manipulate the material world even more efficiently and precisely, raising the wealth of their societies to a similar level as that of their New World counterparts. It would set Eurasia even farther along the path to the industrial age.

Chapter 10

CARVING OUT EMPIRES

One morning in 2001 a group of the feared FARC guerrillas, who were stationed in the rainforests of the southern Columbian region of Amazonas, were astonished to discover that they had been burgled. The thieves had taken several pots, knives, and machetes; and two motors and a generator had been dismantled, and the screws and nuts had all been removed. This was not the only theft in the region. The villages of many of the indigenous tribes had also been raided in the same way. The culprits of these bizarre thefts turned out to be members of the Yuri and Passé tribes, uncontacted horticulturalists and hunters, who had long shunned the outside world and with good reason. Any contact with miners, loggers, farmers, terrorists, or even assimilated tribes would risk disaster: their forest homes could be destroyed; they could be attacked and shot; and, most deadly of all, they could be infected by the diseases of civilization. Dealing with outsiders could result in their annihilation. So it is a mark of the huge value they placed on metal artifacts that they were willing to risk all to obtain them. And the Yuris and Passés are not alone in their high regard for metal objects. Metal knives and machetes were the most common and most valued trade goods that Western colonists used to barter with indigenous peoples, who enthusiastically adopted them. Both the Hadza and Tsimane people we met in chapter 8, for instance, use machetes to clear vegetation and cut up their food, though otherwise they continue to lead their traditional "stone age" lifestyles.

Elsewhere in the world, people have had a longer relationship with metals, particularly in Eurasia, where smiths were crafting copper into jewelry as long ago as the start of the Neolithic. Copper beads dating back to 6000 BCE have been found at the early farming village of Catalhöyük, for example, and in Anatolia smiths soon learned how to beat native copper, silver, and gold into shape. But outcrops of pure metal are rare, so not until around 5000 BCE did large numbers of copper artifacts, not only jewelry but also bladed tools, start to show up in the graves of ruling elites in the fast-growing settlements of Anatolia, Bulgaria, and Mesopotamia. The rise was due to the invention of metallurgy, as smiths learned how to smelt copper from its ores, an advance that led to the rapid expansion in the production and use of this metal. Metal artifacts started to become such common grave goods that archeologists from Christian Thomsen on have used them as an excellent way of dating archeological sites. Archeologists use the presence of particular metals to split the periods that followed the development of smelting into three of the "ages of man": the Copper Age or Chalcolithic, which lasted from around 5000 to 3500 BCE; the Bronze Age, which ran from 3500 to 1200 BCE; and the Iron Age, which started around 1200 to 1000 BCE.

Since the rise of metalwork coincides with increasing wealth around western Asia, with expanding cities, and with the construction of the first palaces, temples, and pyramids, archeologists and historians alike have assumed that metallurgy must have played a vital part. But though they wax lyrical about the beauty of copper jewelry and the craftsmanship of metal weapons, archeologists and historians have been surprisingly vague about what it is that made metals so important; they give no real explanation of how or why the mastery of metallurgy enriched the societies who possessed it. Any world history has to do this and more: it has to explain why metals are so useful; it has to describe how metal smelting emerged and explain why smiths first achieved this difficult alchemy in western Asia rather than elsewhere; it has to show why metal tools were so much better than the stone tools that preceded them; and it has to examine the economic and social consequences of the new technology. As we shall see, metal tools enabled people to further concentrate their mechanical power and use it more precisely and economically. They transformed the ancient world and set Eurasians yet farther along the path of technological progress and dependency on engineering.

The exceptional properties of metals result from their unique atomic structure; they are made up of regular arrays of metal ions surrounded by a sea of electrons. Most metals are reflective and shiny because the electrons are readily excited by photons of light and reemit them when they lose that energy. They also have mechanical properties that are superior in almost all respects compared with those of stone or biological materials. They are stiff and strong because their atoms are arranged in ordered crystalline arrays; to stretch or compress a metal you have to deform the stiff interatomic bonds. But unlike other crystalline materials such as sand or stone, which are brittle, metals are extremely tough because the metal crystals are not perfect but contain many faults and dislocations within and between them. When the metal is heavily loaded, these imperfections relieve stress concentrations, allowing the material to deform and absorb large amounts of energy. This mechanism prevents cracks from running through the metal, so it can resist being stretched as well as being compressed, and it can withstand blows that would shatter materials such as stone or glass. As a consequence, smiths can beat metals into virtually any shape, a property known as malleability, and they can stretch it into long wires, a property known as ductility. They can grind the edge of a piece of metal into a sharp blade, and they can melt it down and pour it into a mold to form almost any shape imaginable.

The main difficulty for the first smiths, though, was how to obtain any copper at all. There certainly were significant deposits of the two main copper ores, the opaque green malachite, and the intense blue azurite, in early centers of cereal farming such as Bulgaria and Anatolia. However, these brightly colored rocks in no way resemble native copper. And their presence cannot be the only reason why metallurgy started in these areas. After all, many other places in Eurasia have much larger deposits of copper ore. The island of Cyprus has such rich ore fields that it was named after the metal, for instance. And even Great Britain has several large deposits of copper ores. The remains of what was once one of the largest Bronze Age copper mines in Europe lie beneath the Great Orme headland in North Wales. In the early nineteenth century, the Duke of Devonshire exploited large copper deposits in Derbyshire and used the proceeds to transform the small town of Buxton, where I used to live, into a major health resort. And in the late nineteenth century, copper produced in the hills of South Wales was smelted and exported in such huge quantities through the port of Swansea that it became known as Copperopolis.

The real reason smelting was invented in western Asia lies in the housing arrangements of people in the region. In the early towns and cities of Anatolia and the Fertile Crescent, people built their homes from sun-dried bricks. These would have been cheap to produce and were easy to lay, but they had one disadvantage: they were not waterproof and were all too prone to being washed away by rain. Furthermore, the only roofing materials available in this region were the trunks and fronds of palm trees, which were strong enough to cover the houses, but would have let in the rain. The solution to both of these problems was lime plaster, which is made by mixing quicklime (CaO) with water to produce a gluey paste. Builders could spread it over the walls and floors and slop it over palm roofs to fill in the gaps, just as medieval builders later used lime mortar to create wattle and daub walls. The plaster would dry out in a few days to produce a smooth, waterproof surface.

The problem was that to produce quicklime, plasterers had to heat limestone to temperatures of over 1500°F (800°C), at which temperature it releases carbon dioxide in the reaction:

$$CaCO_3 \rightarrow CaO + CO_2$$

To do that they needed a new fuel. The answer was charcoal, which is wood that has been partially burned to produce a fuel that is almost pure carbon, but which retains the cellular structure of wood and so has a large surface area. This unique combination enables charcoal to burn at temperatures of well over 1650°F (900°C). To make the new fuel, a charcoal maker, or collier, had to build a solid mound of wood and cover it with soil or turf, before setting fire to it. The covering limited the access of air to the wood, which reduced the temperature of the burn to well below 950°F (510°C). The complex organic compounds in the wood vaporized and broke down, but left the carbon framework of the wood intact. To produce the lime itself, craftsmen had to develop yet another structure: the lime kiln. Early lime kilns were shaped like eggcups: open at the top, so that the lime burners could pour in the raw materials, and with air vents around the base to feed the flames. The lime makers packed alternating layers of limestone and charcoal into the kiln, set fire to it at the base, and let it burn through. After a week the burn was complete, the kiln had cooled down, and they could rake out the lime.

It was just lucky that the conditions within a burning lime kiln were ideal to smelt metals. For if a lime maker placed the beautifully colored copper ores in their charcoal-powered kilns, in the hope of producing a colored plaster or glaze, it would react in a wholly unexpected way. Rather than producing lime that replicated the opaque green of malachite, or the intense blue of azurite, the ores underwent a magical transformation. Since the oxygen supply in the lime kiln was limited, some of the charcoal would undergo only partial combustion, to produce not carbon dioxide (CO_2) but carbon monoxide (CO). This gas could then permeate the ore, producing a reducing atmosphere in which the carbon monoxide would react with copper oxide, removing its oxygen to form the pure metal in the reaction:

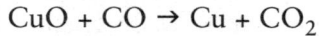

$$CuO + CO \rightarrow Cu + CO_2$$

The liquid metal would fall to the bottom of the kiln, where it would solidify as a lump of copper that smiths could beat into shape or remelt in a pottery kiln and cast in molds fashioned from clay or stone. Seven thousand years ago, the science of metallurgy was born and with it the Copper Age, or Chalcolithic.

All these processes, from mining and transporting the ore, building the kiln, cutting the wood and burning the charcoal, to packing the kiln and shaping the metal, take an awful lot of time, consume a great deal of wood, and require much arduous labor. Metallurgy would only have been worth it if the metal was sufficiently attractive or useful enough so that people were willing to pay the high price it would have commanded. Smiths continued to hammer copper into ornaments such as bracelets and pins for the wealthy, and they also poured it into molds to produce prestige weapons such as knives and daggers for the ruling elite. However, as the technology and trade links improved, and copper became cheaper and more common, smiths also started to make use of its superior mechanical properties to make a whole new range of tools that more people could afford.

People did not immediately stop using stone tools. After all, as we saw in the first part of this book, it is easy to knap razor-sharp stone flakes, and craftsmen can readily haft them onto wooden handles to produce a wide range of cutting tools. Most people therefore carried on using stone sickles and knives. The Iceman, Ötzi, for instance, who died around 3300 BCE and was discovered in a glacier in South Tyrol, Italy, had a flint knife in his rucksack. That Ötzi also had a copper ax, however, points to the most important use of metal tools: to cut wood.

The biggest difficulty with wood has always been how to cut it across the grain. You have to cut through the wood cells, which is why it takes a hundred times more energy than splitting wood along the grain. And if you try to sweep a thick stone ax through the wood, it also compresses the material on either side of the blade, a near-impossible task, and the blade sticks fast in the kerf. Consequently Neolithic wood-workers found it difficult to construct precise joints. The big advantage of the new metal tools is that they could be made much narrower, so they could cut across the grain much more effectively.

Even the earliest copper axes were far better at felling trees than ground stone axes. A study by James Mathieu of the University of Pennsylvania found that a forester using a metal ax can clear wood-land at twice the speed of someone using a stone ax. And because the lumberjack can swing a metal ax more directly at the tree, it wastes far less wood. But Chalcolithic smiths also invented a host of new cutting tools to shape wood once trees had been felled, tools that were used by a new class of specialist workmen: carpenters. Two of the new tools enabled carpenters to smooth and shape wood more precisely along the grain. Drawknives are essentially metal knives with handles at both ends, which a carpenter can sweep sideways along the grain, shaving off the wood. Planes work in the same way, but the blade is held within a framework to limit the depth of its cut, giving more precision. To cut across the grain, smiths made metal chisels that were far stronger and had much narrower, sharper blades than their stone counterparts. Carpenters used them to cut precise grooves into timbers so that they could fix them together into frameworks using strong joints, such as mortise and tenons, and dovetails.

With the saw, which appeared later than the other tools, around 1500 BCE, the carpenter's toolbox was complete. It was the first tool that could cut cleanly and precisely across the grain, and in its distinc-tive operation it was far more effective than a knife. Cutting with the two tools appears at first glance similar. You sweep both of them back and forth across the material using your arm as a crank. However, the cutting actions of the two tools are very different. In a knife, the blade incises a groove through the wood, but even a thin blade quickly jams up against the side of the cut. Knives are therefore only useful in shaving off the surface of small wooden objects, known as whittling. The saw, in contrast, is furnished with large numbers of miniature blades that shave off thin layers of wood at the bottom and sides of the kerf and sweep the debris away. This creates a groove in the wood through which the saw blade moves without getting stuck.

The only downside to copper tools was that they quickly got blunted and snagged because pure copper is relatively soft; the dislocations can all too easily run through the metal. Not until around 3500 BCE did metalworkers learn how to stiffen copper by mixing or alloying it with small amounts of other metals, whose atoms effectively blocked the movements of the dislocations. Smiths first added arsenic, and later tin, to produce the substance we know today as bronze. Bronze is twice as stiff as copper and has the added advantage that it melts at a lower temperature, so it is easier to mold into shape. It became the material of choice for both weapons and woodworking tools for over two thousand years. Its only disadvantage was that copper and tin are both rare materials, and their ores are hardly ever found in the same place. In Europe the main ore of tin, cassiterite, is found in Britain, Germany, and Spain, thousands of miles from the centers of copper production in modern Turkey, Bulgaria, and Cyprus. The two metals had therefore to be transported long distances down rivers or across seas to be combined to make bronze, and the long supply chain meant that it never became a cheap material.

A further, even more important breakthrough, therefore, was the discovery around 1200 BCE by European metallurgists that they could use the same basic technique that they had long used to produce copper to smelt iron. This metal is much stiffer, stronger, and tougher even than bronze, and it had another advantage: its ores are much more common and can be found throughout the world. The only difficulty was that iron has a higher melting point than copper, so early ironworkers were unable to melt it and cast it in molds. To heat iron up enough to smelt it, smiths had to mechanically ventilate their kilns using bellows, and they had to work the iron mechanically, hitting the soft bloom they produced with metal hammers, to harden it and beat it into shape. Iron therefore had higher labor costs than bronze, but it could be produced just about anywhere. Its greater availability and reduced transport costs meant that it became significantly cheaper, and in time it could be made into much better, harder-wearing woodworking tools.

At first glance it is hard to see how improving woodworking tools could have advanced the lot of mankind or changed society. Obviously, it would reduce the time and energy that craftsmen needed to shape wood, making their products cheaper, and allowing them to produce more finely shaped furniture, but this would hardly transform people's lives. However, as we have already seen, novel tools can make new technologies practical for the first time and open up new worlds. The hand ax, for instance, enabled *Homo erectus* to build wooden shelters and carve large spears and so enabled them to live permanently on the ground, while the ground-stone ax enabled the LBK people to clear the northern forests and spread their farming lifestyle across Europe. Metal woodworking tools changed the world in even more dramatic ways. They allowed craftsmen to develop two new forms of transport that would transform the infrastructure and economics of the Old World: wheeled vehicles and plank ships.

The first part of this book showed that our ability to carry our tools and our food about with us was from the start one of the keys to our success. But there is only so much a person can carry, typically around fifty pounds (23 kilograms). And a lot of the energy we use to carry an object about is wasted simply supporting its weight, whether we hold it in our hands, in a rucksack, or on top of our heads. Herders learned how to reduce this cost by loading their goods onto travois and dragging them along the ground. The cereal farmers of Mesopotamia developed an even more efficient means of transport, the sled. The technology is simple—sleds are simply wooden boxes fitted with smooth runners to reduce friction—but they are surprisingly effective, even without snow to coat and lubricate the soil. Sleds were being used as long ago as 4000 BCE in early Mesopotamian cities such as Uruk; they were the main form of transport in Scotland well into the eighteenth century; and they continued to be used by farmers in Ireland up until living memory. Since early cereal farmers were already using oxen to pull their plows, they also started to use their beasts to pull their sleds. And because sleds are easier to pull than plows, the farmers could also employ smaller domesticated ungulates such as asses, donkeys, and horses.

A potentially more efficient way of moving goods is to mount the sled on wheels, circular disks that are attached to the ends of a slender axle, which in turn is housed below the bottom of the vehicle. Wheels lower the power needed to pull a vehicle by reducing friction; the resistance to motion of a wheeled vehicle due to friction around the edge of the axle can be reduced by lubricating it, and in any case friction on the axle has less effect since it acts much closer to the center of rotation than the rim. However, it is not immediately apparent how people invented wheels. Archeologists traditionally assert that the idea of wheels came to people as a result of their attempts to move large stones or statues across country. The tale we have all been told at primary school is that they rolled them on the tops of a series of logs. However, this seems unlikely, since the log-rolling technique is highly impractical.

For a start it would take a huge investment in time and energy to fell, debark, shape, and transport the large number of identical logs required, something that would be difficult using only hand tools. And there would have to be not one but two teams of people to shift the stone: one to pull the sled; the other to pick up logs from the rear of the sled, carry them to the front, and lay them down at precisely the right angle in front of the moving sled. If two of the logs touched each other, they would also jam up the operation since the rear of the log in front would be moving in the opposite direction to the front of the one behind. The process would be stop-start and would not only be hard work but downright dangerous. Besides, all the evidence suggests that people moved heavy stones by dragging simple sleds across country, as is often depicted on the walls of ancient Egyptian tombs. The Egyptians lubricated the movement of their sleds by pouring water in front of their runners, while the builders of Neolithic monuments such as Stonehenge and the Ring of Brodgar would have used animal fats or seaweeds as lubricants.

It is far more likely that the idea of wheels did not come top down from major infrastructure projects, but bottom up from people's homes. As we saw in chapter 7, people had been using small rotating devices, drills and spindles, for thousands of years. They had also fitted them with circular stones—whorls—and had learned how to restrict the motion of pump drills by making the axles rotate within holes. They would quickly have realized that if they laid the arrangement on its side and added a second whorl, they could roll it across the beaten-earth floor of their dwellings. Model animals fitted with two pairs of wheels would have made ideal toys for their children. And such toys have indeed been found in archeological sites around the world. They were even made in the New World, where full-size wheels were extremely rare; the Mayans, Aztecs, and Incas fashioned toy jaguars, monkeys, dogs, and even alligators equipped with four clay wheels.

The only problem was how to scale up model wheels into full-size equivalents. Unfortunately, without saws you cannot slice a tree up into circular segments, and even if you could, a pie-shaped groove would open up as the disk dried out and it would fall apart. The only way to make a wooden wheel with stone tools would be to hack away at a tree trunk until you produced an object like two cones attached base to base. You could then drill through it from point to point and attach the axle in the hole. Such a wheel would work, but it would be extremely heavy and would be limited in diameter to the width of a tree trunk. Small models and pictures of wheeled vehicles using this design have been found in the Middle East. They date from the early fourth millennium BCE, while much later Aztec ones have been found in Mexico. However, the Aztecs, who were still using stone tools, took the technology no further, while from the early Bronze Age onward, people in the Caucasus and Fertile Crescent found that with the advent of metal tools they could make much larger, lighter wheels.

Bronze Age wheelwrights used a simple but effective method. First, they split wood into planks and laid them side by side, carved them around the edges into a circle, and attached them to each other using precise tongue-and-groove joints. Finally, they chiseled grooves at right angles across the planks and inserted batons to strengthen the wheel across the joints, fixing the battens with metal brackets or nails. The design wasn't perfect but it worked, and wheeled vehicles became common in the second half of the fourth millennium BCE, particularly near the centers of metallurgy in the Balkans and the Caucasus. The earliest pictorial evidence of wheels is on the Bronocice pot from mid-fourth millennium BCE Poland. It is in the form of circles drawn at the four corners of the symbol for a sled, suggesting that it represents a four-wheeled wagon. The earliest surviving wheel, meanwhile, is the Ljublyana Marshes wheel, from Slovenia, which dates to 3200 BCE. The first wheeled vehicles were probably two-wheeled handcarts, which go round bends easily, but less maneuverable but larger four-wheeled wagons soon followed, and like plows and sleds, they were soon being harnessed to oxen.

Wheels spread rapidly across Europe and Asia, along with the first purpose-built roads. The earliest paved highways appear to have been built in Uruk as early as 4000 BCE; corrugated roads made using lines of split planks set at right angles to the road reached Glastonbury,

England, by 3000 BCE, and tramlines with their rails set parallel to the road reached Ireland by 2500 BCE. The earliest surviving wagons were found on the Russian steppes, in burials dating from around 2000 BCE. Wherever wheeled vehicles were produced, people found ways of harnessing the local type of ungulate to pull them: oxen in Europe, asses in the Fertile Crescent, and horses on the Russian steppes. And as woodworking tools improved, craftsmen developed better ways of making larger, lighter, and stronger wheels. They joined curved lengths of wood, or fellies, to make a rim and connected them to the central axle via wooden spokes. By 1500 BCE wheelwrights had perfected the design of the wooden spoked wheel, which is still used today.

But wheels have their limits. First and foremost, they only run smoothly on hard, smooth ground; they are useless in mountainous terrains, in wet boggy areas, and in sandy deserts. And they demand roads that have to be built and maintained. Consequently, until the eighteenth century, wheeled haulage vehicles were only used for moving goods short distances—around the farm, to and from market, or around town. And coaches were slow, uncomfortable vehicles, so in town many people preferred to be carried in sedan chairs. Another more efficient form of transport was needed to expedite long-distance haulage, and a more effective way of reducing the cost of moving goods—and people found that this was best done on water.

Paleolithic and Mesolithic hunter-gatherers and Neolithic traders had long been able to build useful watercraft using their stone tools. They hollowed out tree trunks to produce log boats, tied reeds together into rafts, and covered wooden frameworks to produce skin boats. These small craft would have been good enough to transport people and their goods along rivers and around coastal margins. However, none of these sorts of boats would have been large or strong enough to make them seaworthy so that they could transport large quantities of metals reliably from their place of origin to the workshops of the growing civilizations of the Middle East, Europe, and Asia: tin from

Britain and western Europe and copper from Cyprus. Nor could they satisfy the demand from the ruling elites of the growing empires to transport a wider range of both basic and luxury goods: huge stones and timbers to construct their palaces and temples; fabrics, ornaments, jewels, ceramics, luxury foods, and spices.

The answer was to construct larger vessels—ships—by fastening planks of timber together side by side to produce a watertight shell that could be stiffened with an internal framework so that it could withstand the huge bending forces that a rough sea would impose on it. There are various ways of doing this. You can join the planks together first, then insert the frame, or you can build the frame first and attach the planks. You can butt the planks side to side to produce a smooth hull, or you can overlap them to produce a stepped "clinker" hull. But all forms of shipbuilding demand even greater precision in the cutting and shaping of the wooden members than you need to construct wheels. A shipwright has to take particular care when they cut and shave the individual planks of wood, or strakes, and has to devise more and more elaborate ways of joining the strakes perfectly to prevent leaks. And because wood shrinks and swells as it loses and takes up water, and the bending forces on a ship battered by waves tend to shift the planks relative to each other, shipwrights also had to develop different methods to plug the tiny gaps that inevitably opened up between the timbers. They used caulking made of plant matter, bits of old rope, and bitumen.

Given the huge skill and precision required to build them, it is not surprising that plank ships only appeared early in the Bronze Age, in the second half of the fourth millennium BCE. The oldest surviving craft is the funeral vessel of the Egyptian pharaoh Khufu the Great. This ship was built around 2500 BCE before being disassembled and stored under the Great Pyramid of Giza, presumably ready to be rebuilt in the afterlife to carry the dead king across to the underworld. This ship was already highly sophisticated in its structural design, being constructed from planks of cedarwood, which were fixed securely together using complex mortise-and-tenon joints. This suggests that plank ships had long been used for more corporeal purposes: for importing cedar timbers from Lebanon, for instance. And there is further evidence that shipwrights and sailors had even developed a method of overcoming the main difficulty of water transport—powering a vessel that weighs many tons—over a thousand years before the pyramids were built.

As the British engineer Isambard Kingdom Brunel observed when he was planning the third and last of his giant ships, the SS *Great Eastern*, in the mid-nineteenth century, the larger a ship, the more efficient it will be at transporting goods; the volume of goods it can hold increases with the cube of its dimensions, whereas the resistance of the water to its motion increases only with the square of its dimensions. The problem that he experienced is that bigger ships still need more power than small ones, and the steam engines of his day were not powerful enough to provide it. The builders of the first plank ships had a similar problem. Early boats were mostly paddled or rowed, but as ships got larger, they needed longer and longer oars and more and more men to pull them. Mariners could force galley slaves to row, but these people had to be fed and watered, making long-distance transport less profitable. Fortunately, on one of the first great trading routes, the River Nile, the conditions were ideal to develop a novel way of powering ships—sails—to harness for the first time one of our major resources of renewable energy: the wind.

A sail seems to be an obvious solution for how to power ships. After all, open seas are particularly windy places, and holding a large piece of cloth aloft is the ideal way to increase the propulsion force. Unfortunately, though, simple square sails mostly increase the drag force on a boat; they can only be used to sail a ship downwind. Fortunately, this was not a deal-breaker in Egypt because of its geography and climate. The main river, and the source of all its fertility, was and remains the Nile, which runs almost exactly from south to north. The current of the river was sufficient to float the huge stones the Egyptians needed to construct their pyramids, temples, and statues, from the limestone and granite quarries of Aswan, high upstream, down to the main cities of Giza and Thebes. To return the ships upstream, the sailors could make use of the northeasterly trade winds. Simply by hoisting a square sail up on an A-frame mast, they could sail the ship back upstream at all times of year except during the summer floods. The first image of a sailing ship is on a piece of Egyptian pottery that dates right back to 3500 BCE.

Egypt soon became something of a backwater in shipbuilding, however, since there are few suitable forest trees in the country to provide the timbers, but the giant size of Khufu's ship—it was 160 feet (49 meters) long—shows that progress in ship design must have been rapid. Later in the Bronze Age the Egyptians probably relied on ships built across the Mediterranean, in particular by the Phoenicians of their vassal cities Tyre and Sidon, in present-day Lebanon. Unfortunately because of the harsh conditions of the Mediterranean, little physical evidence remains of Phoenician vessels in the form of shipwrecks, but discoveries of the cargoes that survived shipwrecks and images of the ships themselves show that maritime trade quickly grew. Bronze Age trade routes linked Egypt with Minoan Crete, Mycenaean Greece, and the copper mines of Cyprus and even extended into the Atlantic Ocean to the tin mines of Cornwall and into the Persian Gulf to link up with the civilizations of Mesopotamia and the Indus Valley. In ancient times marine architecture reached its apotheosis in the giant grain ships that were one of the keys to the longevity of the Roman Empire. Reaching over two hundred feet (60 meters) in length and weighing over two thousand tons, these double-masted giants carried grain from the breadbasket of Egypt two thousand miles (3,200 kilometers) across the Mediterranean all the way to Rome to feed the fickle and (as we saw earlier) bloodthirsty inhabitants of the largest city of the ancient world.

By the end of the Bronze Age, the smiths who made the new metal tools, and the carpenters who used them to build the new transport systems, had transformed trade, allowing farmers to take more goods to market and merchants to trade goods farther and faster across the water. As trade flourished, the wealth of Bronze Age societies increased, but it also led to a further rise in inequality. Because of the high cost of bronze only a small ruling warrior elite could enjoy the benefits it brought. There is no doubt that at this stage in the development of civilization, ships were far more important and influential vehicles than wheeled carts or wagons ever were, which is why major cities—then as now—were ports, situated on the coasts or beside a major river, and why states could grow ever larger. However, as we shall see in the next chapter, in the long run, the concept of the wheel was to have an even greater influence on human society and technology because of the way it was incorporated into machinery.

Chapter 11

GEARING UP

If you want to see the apotheosis of industry in the ancient world, the place to visit is the Barbegal mill complex, the remains of which can be found on a steep escarpment eight miles (13 kilometers) east of Arles in the Camargue region of southern France. At the end of a fine Roman aqueduct and running down the slope are the remains of two water channels and the foundations of sixteen overshot waterwheels, in two sets of eight. Despite its obscure location, the complex is larger than the one on the Janiculum in the Eternal City itself; archeologists have estimated that it could have developed around forty horsepower (thirty kilowatts) and ground over four tons of flour per day, five hundred times faster than a person using a saddle quern. It could have produced enough flour to supply the local town of Arelate. This early example of an industrial complex was the largest and most spectacular of the laborsaving machines that people had started to develop from the Bronze Age onward and that transformed premodern industry. All of them were based on the simple concept of the wheel, but as we shall see, the journey to develop and perfect machinery that was to culminate in the Barbegal mill complex was long and circuitous.

Despite their many advantages, even the sophisticated metal tools wielded by ancient craftsmen—axes, hammers, saws, planes, and knives—have their drawbacks. They need to be swung back and forth by the user, who therefore has to expend a great deal of energy to repeatedly accelerate and decelerate not only their tool but also their hands, arms, and sometimes their whole body. Most of this energy is wasted, so a craftsman's productivity is limited by the low long-term power output of the human body, just eighty watts; we are simply not adapted to perform strenuous repetitive motions for hours on end. Another drawback of powering tools by swinging them back and forth is that it is difficult to train the head of the tool in precisely the same motion stroke after stroke. Hand tools are therefore inevitably imprecise.

Using wheels mounted on stationary axles can overcome both of these problems. As people would have noticed when looking at the motion of spinning tops, or at the rotation of the wheels of upturned carts, circular motion has almost magical properties. Though each point on a wheel may move rapidly to and from the same place, the motion requires little energy to maintain it because it is always accelerating at right angles to its velocity. The only power required to keep a wheel moving is to overcome the friction at its axle. The other unique aspect of a wheel is that each point on its rim returns to *exactly* the same position at each rotation, which is why the drills we examined in chapter 7 cut perfectly circular holes. Rotation is therefore a key to precision engineering. Once craftsmen had learned how to mount wheels in solid housings, they were able to invent machinery that could shape a wider range of artifacts more quickly, more perfectly, and using less power than traditional hand tools.

The first of these simple rotating machines was the potter's wheel, which first appeared soon after the earliest wheeled vehicles, in the second half of the fourth millennium BCE. Early potters would have found that shaping soft clay by hand into watertight pots was difficult. One of the first and most efficient techniques they developed was the coiling method. The potter first rolled the clay into a series of long thin ropes before building their pots bottom upward by coiling the ropes together into a close-bound helix. They could then smooth the coils together inside and out to create flat surfaces. Unfortunately this was a lengthy process, made all the more difficult because the potter continually had to turn the pot or move around it to get at all the sides. The first potter's wheels were merely turntables, but potters gradually realized that if they could spin a wheel around at high speeds, they could use the energy in the spinning table to shape the clay into perfectly circular containers. At first they had to rotate their tables by hand, but over the centuries they developed taller tables and mounted them on longer axles with a platform around the bottom, so they could rotate the wheel using their feet, freeing both hands. By 3000 BCE, the potter's wheel as we know it today had been perfected. Once they had thrown the pot, craftsmen could also use the wheel to decorate it, by dipping a brush into glaze and holding it against the pot as it spun around, or inscribe it with circular grooves using a knife.

The method of incising grooves in pottery was probably an inspiration for the development of a second machine, which has proved even more important than the potter's wheel: the lathe. The advent of iron smelting around 1200 BCE enabled blacksmiths to produce chisels and knives that were hard enough to cut into wood without wearing down too fast. To create finely shaped wooden objects with perfectly circular cross sections, craftsmen needed simply to clamp blocks of timber at their two ends via smooth iron bearings onto a rigid frame, so that the craftsmen could spin the blocks around like the axles of a cart. The lathe operator could then hold the point of a chisel up against the rotating wood to gradually peel material off and create a whole range of circular objects: tool handles, the axles and spokes of wheels, the legs of chairs and tables, and vessels such as cups, bowls, and plates. Wood turning quickly became a profession, with several specialties, from the bodgers who worked in woodlands to produce simple poles to town-based craftsmen, such as the cup makers who worked in Viking York's Coppergate, or Cup Street. Until the eighteenth century, when vast factories started to make cheap, colorful pottery, most people ate and drank off wooden plates, bowls, and cups. They were not only cheaper but were far more robust than china ones and could last people their whole lifetime.

Egyptian craftsmen operating a lathe. The man on the left rotates the piece using a bow, while the man on the right is cutting away the pattern with a chisel.

A rotating wheel can itself be used as a shaping tool. Craftsmen realized that if they made a wheel out of a hard, abrasive substance such as stone, they could hold a wooden or metal object up against it and gradually grind it away. They could smooth wood and sharpen the blades of metal tools such as knives and hoes. Sanding and grinding machines using this principle were cheap to make and easy to use, so traveling knife grinders were common street traders right up until the middle of the twentieth century, roaming residential areas to hawk their services. And if smiths forged a thin metal disk and cut teeth into its edge, they could use it as a cutting tool. As iron and eventually steel became available, the smiths developed circular saws that could readily cut through wood.

But perhaps the most important rotating device in the ancient world was the one that people developed to power one of the most labor-intensive task of all in cereal farming: milling grain. We saw in chapter 8 how time-consuming and backbreaking saddle querns were to use and what a toll they took on women's bodies. The solution that reduced at least some of the workload was the rotary quern, a tool that, like the saddle quern, was made from two heavy stones. In the rotary quern, however, the upper stone was shaped like a ring doughnut and was rotated above the lower one using a wooden handle inserted near its rim. It was held in place by a projection from the center of the lower stone that acted as an axle. Women could pour grain into the quern through the central hole and rotate the upper stone, grinding the grain, which worked its way outward as it was progressively crushed. Experimental archeologists have shown that rotary querns can process grain three times as fast as a saddle quern, trebling productivity from two to six pounds per hour (1–3 kilograms per hour). The rotary quern was invented around 500 BCE somewhere in the Middle East, and quickly made its way across Asia and Europe, reaching Highland Scotland by around 200 BCE, where it continued to be used until modern times.

But even though rotating machines are more efficient than reciprocating ones, they do still need to be driven around to overcome friction and power their operations. The commonest way to do this was to attach a handle and push them around, using the arm as a crank. The user provided the energy with their shoulder muscles, swinging the upper arm back and forth, and allowing their elbow to bend passively to allow the hand to sweep around in a circle. This works well for grinding grain, but since it can only move wheels around relatively slowly—twice a second at best—it is nowhere near fast enough to power a lathe. Early Egyptian turners speeded up the rotation of their lathes by using a bow to spin the axle or employed a boy to wrap a belt around the axle of the lathe and spin it by pulling the belt back and forth.

An even better way of driving rotating devices, while keeping both hands free and minimizing labor costs, was to use the operator's feet to operate a treadle. One machine that used a simple treadle system was the medieval pole lathe. Pushing down on the treadle pulled down on a cord that was wrapped around the axle of the lathe, spinning the piece of wood. The other end of the cord, meanwhile, reached up to the end of a long springy pole and bent it downward. Releasing the treadle allowed the pole to straighten again and the piece to spin backward. Regularly pressing down on the treadle could therefore set up a reciprocating action in which energy was transferred between kinetic energy in the rotating lathe and elastic energy stored in the bent pole. An even better arrangement was to attach a rigid crank, rather than a rope, between the treadle and the wheel, an arrangement that was incorporated into the traditional spinning wheel. Pushing the treadle rhythmically up and down can produce continuous rotation, especially if the flywheel is heavy enough to store sufficient amounts of rotational kinetic energy.

The new rotating machinery provided the precision and power needed for a whole range of manufacturing processes and raised a worker's productivity, but it did not totally banish the muscular power and energy requirement. Paradoxically, that final stage was facilitated by developments in the techniques that engineers developed to help with the most energy-intensive task that early farmers needed to perform: transporting water.

Much of the land into which the early cereal farmers had expanded—the Fertile Crescent, Egypt, and the Indus Valley—was watered not by rainfall, but directly from the major rivers that flowed through it. Farmers irrigated their crops by digging a series of canals, ditches, and dams that diverted the water away from the river and across the wide area of the floodplain. The floodwaters also brought nutrient-rich silt that fertilized plants so farmers could harvest more than one crop year after year without impoverishing the soil. By the end of the Bronze Age, for instance, the soils of Egypt, replenished each year by the Nile, supported a population density of two hundred to three hundred per square mile (80–120 per square kilometer). However, as the population in these areas grew, farmers started to cultivate land that lay higher up and farther away from the main stream. They dug new ditches through the soil to direct the water, but they also had to lift the water to fill them and irrigate the land.

Unfortunately, the human body is poorly adapted for repetitive and arduous lifting tasks such as raising water. You have to bend down and lower your container into the water, let it fill, then straighten up and lift the container and its charge of water before emptying it into the higher trough. As anyone who has done a lot of lifting knows, it is strenuous and dangerous. Much of the energy you use is wasted moving your body and the container, and the movement puts a huge strain on the human back, which did not evolve to straighten up against large loads; mine has never been the same since I helped move a piano down four flight of stairs back in the 1980s. These problems prompted the farmers of the Fertile Crescent to develop the first lifting device—the shadoof. As a machine the shadoof is extremely simple: it consists merely of a pole hinged a fifth of the way along to a supporting framework, with a rope and bucket attached to the long end of the pole and a heavy counterweight attached at the other. The counterweight is just large enough to raise a bucketful of water from one body of water to another a yard or two higher up without any effort on the part of the operator. To power the device, the user only has pull the bucket down once they have emptied it. Pulling downward with our body and arms like this is a much more natural action, since it involves much the same actions that we perform when we hit or throw something.

The shadoof has an overall efficiency of over 60 percent and is such an effective tool that it is still used in the Middle East. However, it has to be worked by hand, so raising water with it still uses up a lot of human energy. In the Iron Age, farmers therefore started to develop a whole new range of wheel-based machines that they could power using their draft animals. The saqiyah consists of a line of buckets on a rope that goes around the rim of a wheel, or sometimes two wheels, one mounted above the other, with the rope acting like a bicycle chain. Rotating the upper wheel caused each bucket in turn to pick up water at the bottom of its travel, lift it, and pour it out into a channel or trough at the top of its travel. The first saqiyahs were probably hand cranked, but people realized that if they linked the upper wheel via pegs to a second, horizontal wheel, the first proper gears in history, they could harness an ox, donkey, or camel to it via a pole and set the animal walking in a perpetual circle. The circular treadmill was born and was soon used to power a series of increasingly elaborate and efficient lifting devices.

In the tympanum, the buckets were replaced by a single hollow drum, separated into eight radial compartments, each of which had one opening on the rim and a second at the axle. As the wheel rotated, water entered each segment when its opening was underwater and was then lifted until the bottom wall of the segment reached horizontal. At this point the water moved back inward and through the hole at the axle into a trough just below it. Having much larger containers than the saqiyah, the tympanum could raise larger amounts of water, but it could only raise the water by around a third of the height of the wheel, and it was only 50 percent efficient, since the water sloshed rapidly inward at the top of its travel. The scoop wheel, which was invented in Egypt around 500 BCE, overcame this problem. It had a large number of curved buckets that raised the water and moved it more steadily to the center of the wheel, thus increasing its efficiency to over 60 percent. Most efficient of all was the Archimedes' screw—a tube split into a single helical chamber, which was mounted at a small incline with its mouth submerged into the lower water trough. As it was rotated, it swallowed, lifted, and drew water steadily up its length, before releasing it at the other end into a higher trough. Archimedes' screws reached efficiencies of well over 70 percent and were so effective that they are still used in modern hydraulic-engineering projects. At the Tees Barrage in northeast England for instance, they are rotated to raise water in times of drought, and when the river is in spate, they harness the flow of water to produce electricity.

To reduce energy losses caused by friction in the gears, the ancients also developed a more efficient method of harnessing the power of their animals: the treadwheel. Rather than having an animal walking in a circle around a horizontal wheel, the treadwheel is a huge vertical wheel with open sides and a horizontal axle—like a gigantic hamster wheel—in which a draft animal can be set walking. Treadwheels are rarely used today, but if you visit Carisbrooke Castle on the Isle of Wight in England, you can enjoy the charming sight of the resident donkeys reenacting the way the castle's defenders used to raise their drinking water from the well. Treadwheels are certainly efficient, but the animals need a lot more training to use them, which is why engineers often modified them so that they could be powered by workmen or slaves. The Romans used treadwheels to power their cranes, and medieval masons mounted them high in the walls of cathedrals when building them to lift the heavy stones and roof timbers. I have seen a splendid working example in the roof of Beverley Minster near my home in East Yorkshire, England.

But though draft animals took the physical burden of lifting water away from people, the beasts themselves needed food and water. So people were always on the lookout for a way to lift water without using any muscular effort at all, and in fourth century BCE Egypt, engineers found one: they could use the flow of the River Nile not just as the source of irrigation water, but as the source of the power to lift it as well. They came up with the noria, which looked just like a saqiyah, but which not only had buckets around the side of its rim but a series of paddles on the outside of the rim. The flow of the river pulled the wheel around, like in an undershot waterwheel, and the buckets lifted water to a trough at the top of the wheel. Norias quickly spread throughout the Middle East, were commonly used in China, and are still in use today, notably on the River Orontes in Hama, Syria, where there are some majestic examples, some sixty-nine feet (21 meters) in diameter, which were built in 1361.

The impact of water-lifting devices in the irrigated farmlands of the Middle East, the Indus Valley, and southern China was profound. Being able to raise water so effectively enabled farmers to cultivate a much larger area. In Egypt, the population density doubled during the Iron Age, reaching 570 per square mile (220 per square kilometer) in late antiquity, comparable to that of the Aztec capital Tenochtitlán. Elsewhere, the impact of forced irrigation was less benign. In Mesopotamia, irrigation water brought in salts that were deposited in the soil as it evaporated, resulting in a sharp rise in soil salinity. Crop yields almost halved between 2400 and 2100 BCE, and farmers were forced to grow more salt-tolerant but lower-yielding barley. The population, wealth, and power of the region consequently declined, and nowadays most of the Fertile Crescent is arid and infertile, possibly the first example of a man-made environmental disaster in history.

Further north, in Europe, farmers did not need to use machines to raise water because the rainfall is higher and supplies cereal crops with all the water they need. However, the Romans were quick to realize that they could reverse the action of saqiyahs and use the flow of streams to provide power for milling grain. They developed several types of waterwheel, devices that were described by the first century BCE engineer and writer Vitruvius. Probably the first devices they built were undershot wheels, like the ones on the Janiculum in Rome, which were essentially barely modified norias. But the Romans also developed other more efficient ones. In the lowlands they built overshot wheels, like the ones at Barbegal, while in more mountainous areas they built vertical-axis waterwheels. In these devices, they diverted water through a sloping pipe and directed it against angled blades that were arranged around the edge of a horizontal wheel. The impulse of the water pushed the apparatus around, just as the wind causes the sails of a windmill to rotate, driving a quern stone without the need for any gearing.

By the end of the classical period, the ingenuity of engineers had enabled them to develop most of the rotating devices that would later be used to power the Industrial Revolution, some twelve hundred years later. But that we have found so few examples of large machines from ancient archeological sites shows that the ancients had made little progress in developing an industrial, machine-based society. They still relied for the most part on human muscle power to provide the mechanical energy to grow and process food, build their houses, and make their furniture. In a world with plenty of slaves to provide the labor, there was little incentive to develop energy-saving devices. As we shall see, it would be over a thousand years before people found it worthwhile to make machines that could develop more than three to four horsepower (2–3 kilowatts) and make them in large enough numbers so that they could take over from workmen and draft animals. And before that happened another technological revolution had enabled people to harness chemical energy to kill each other in larger numbers and with much less effort than ever before, transforming warfare.

Chapter 12

FORGING MILITARY POWER

When I was at school, I was fortunate to learn some ancient history. Having taken our O-level Latin examination a year early, a small group of us were taught by a former army captain about ancient Greece. It was marvelous fun. We touched on Greek philosophy and Athenian democracy, of course; the building of the Parthenon; the great tragedies of Aeschylus, Sophocles. and Euripides; and the bawdy comedies of Aristophanes. But most of all we learned about wars: about Themistocles, Miltiades, and the defeat of the Persian invasions of Greece; about Pericles and Alcibiades and their role in the ruin of Athens; and about how Philip of Macedon and his son Alexander the Great later defeated the other Greek city-states and went on to conquer the vast Persian Empire. Our teacher waxed lyrical about battles on sea and land: about triremes, hoplites, phalanxes, and the oblique battle line. I was even persuaded to buy copies of the two earliest history books: Herodotus's *Histories* and Thucydides's *History of the Peloponnesian War.* Both books were packed with detail, and Herodotus's work was full of great stories, some of them true.

But I gradually started to realize that, like the histories that followed, they failed to answer one fundamental question. Why did these wars take place at all? Why would people leave their homes to kill large numbers of their fellow human beings and destroy their livelihoods? In particular the quarrels between the Athenians and the Spartans that were the ostensible causes of the Peloponnesian War hardly seemed like justification for the thirty years of mass slaughter that followed. Their disputes seemed to be more like the "he said, she said" squabbles of small children; with a few sensible compromises, they could easily have been settled. Nor did the histories even begin to explain the larger trends in history. Why did certain powers emerge, prosper, and decline? What determined the outcome of wars between states? The traditional assumption, that it was all down to the brilliant vision and leadership of "great men," hardly seemed like an answer. History looked like a bewildering succession of pointless and destructive struggles.

We have to look to primatologists and anthropologists to explain why humans engage in such counterproductive activity. They have shown that we can learn a lot about ourselves by examining the behavior of our closest relatives, the great apes. And it is possible to identify trends in history that are related to a state's agricultural productivity and the technological capabilities of its craftsmen. Military power has less to do with the personalities of leaders and more with the ability of a society to marshal and concentrate physical power. Consequently, if we know something about the technological developments that took place in Eurasia following the beginning of cereal farming, we can start to make sense of the rise and fall of empires in ancient times, and the interminable series of wars that plagued the region in the premodern period. We can start to see the underlying patterns in what we have traditionally been led to think of as history. Ultimately, this approach can explain how a few Spanish adventurers were eventually able to travel to the New World and crush the mighty empires of Central and South America.

In the mid-1970s, the British primatologist Jane Goodall was shocked to observe a conflict between two groups of chimpanzees at her Gombe Stream National Park study area in Tanzania. Following the fission of her main study group, members of the larger faction, the

Kasakele, started to target males of the smaller, breakaway Kahama group. Several male chimps would silently stalk into their enemies' territory and ambush solitary individuals, beating them with their fists, tearing at their limbs, and leaving them to die. Over four years they killed all six males of the smaller group, along with several females, while the two remaining females were beaten and kidnapped. With all the members of the Kahama group gone, the Kasakele took over their territory. The Gombe chimpanzee war was a terrible shock to Goodall, who had previously discovered a host of much more positive human-like traits in the species—that they used tools and cared for their fellow chimps, for instance—so she had started to feel that chimpanzees were kinder than humans and hoped that this violent saga was merely an aberration.

Later studies, however, have replicated Goodall's finding and shown that war is a constant feature of chimpanzee life. From the perspective of natural selection, chimpanzees have good reasons to be kind to fellow members of their group, but to kill members of neighboring groups. Within a small band of animals, selection should tend to favor individuals who help other members of the group because most of them would be relatives and share some of the same genes. In addition, a peaceful, mutually supportive group will survive better and expand in size more rapidly than one beset by competition and riven with internal strife. In contrast, individuals within neighboring groups of animals are unrelated and offer competition for food and territory. Natural selection should therefore favor animals who not only guard their own territory but kill their neighbors if they can do so without fear of being harmed. Hence the aggression of the Gombe chimps, and the ambush tactics they employed to gang up on single individuals of the smaller group; they could kill their lone enemy while minimizing the chances of being killed or injured themselves by their hapless victims.

We have unfortunately inherited this native xenophobia from our ape ancestors, along with our engineering skills. Indeed, the selection pressures on humans to kill members of neighboring groups have been even stronger because we are so much better at doing it. With our superior ability to employ percussive tools as weapons, we can dispatch other people with just a single blow, greatly reducing the danger of being hurt ourselves if we ambush an enemy. So it should come as no surprise that warfare is a common feature of the life of small bands of hunter-gatherers and horticulturalists. The tribes of the New Guinea highlands, and the feared headhunters of Borneo, regularly used to ambush individuals of neighboring groups or make surprise raids on their villages. And just like chimpanzees, the raiders didn't necessarily kill all the people they attacked. Following surprise attacks, they might capture and enslave some victims, particularly women, to add to the size and strength of their own tribe.

We used to consider tribes who indulged in this behavior as primitive or bloodthirsty, but the sad fact is that in "civilized" societies, we kill far more of our enemies in wars than hunter-gatherers ever did. The difference is that we usually pay other people—our armies, navies, and air forces—to do it for us; we celebrate our victories in films rather than in poetry and song as the ancients did; and we honor our own "fallen" with grander monuments and commemorations. And the numbers of people we kill in our wars has risen from single figures to tens, hundreds, thousands, and, in the last century, millions.

In the past fifty years, archeologists have also been finding that warfare goes back deep into prehistory. In 2016, for example, Dr. Marta Mirazón Lahr from the University of Cambridge and her colleagues found evidence of a massacre that had taken place ten thousand years ago on a lakeshore at Nataruk near Lake Turkana in Kenya. They found the unburied remains of twenty-seven hunter-gatherers, including eight women and six children, most of whom showed signs of a violent death. Many skeletons exhibited signs of blunt-force traumas and several had stone projectiles still lodged within their skeleton. They must have been ambushed while out on a foraging trip. Evidence of simmering small-scale disputes go back even further. Analysis of skeletons in a 13,400-year-old cemetery in Jebel Sahaba, Sudan, by Isabelle Crevecoeur at the University of Bordeaux, showed that many of the people had suffered repeated lesions, including impact marks and stone debris from spear- and arrowheads. These findings show that there must have been almost constant and bloody disputes between early hunter-gatherers.

The incidence and scale of conflicts dramatically increased, however, as people took up farming and population density rose, since there would have been more competition for land, and villages held large stores of food that raiders could steal. The first known incidence of large-scale warfare in Neolithic Europe was found at a site at San Juan ante Portam Latinam in northern Spain. Teresa Fernández-Crespo from the University of Oxford and her colleagues found that among the 338 skeletons in a mass grave that dated to between 3400 and 3000 BCE, 45 percent of the adult males had evidence of violent traumas consistent with being wounded by axes or clubs. Some even had arrowheads lodged within their bodies. The group must have been involved in a protracted military struggle.

As the scale of warfare increased, there was also pressure to develop better weaponry. The earliest weapons people used were no doubt the same ones they used for hunting: sticks and stones, clubs and spears. As we saw earlier, by swinging and throwing these sorts of weapons, we can readily impart enough energy—thirty to one hundred joules— to fell a fellow human, break their bones, or perforate their flesh and major body organs, at least if we hit them in the right places. And since Paleolithic people had also developed more sophisticated projectile weapons, powered by slings, spear-throwers, or bows—weapons that they could deploy at greater distances—they could kill people from farther away, which would have been a great advantage in battle.

Indeed, the invention of sling-projected darts may have been one of the factors that enabled modern humans to exterminate the Neanderthals. Following their invention, around fifty thousand years ago, humans could kill individual Neanderthals from a greater distance and be less likely to be drawn into deadly close combat. This story certainly fits in with the timing of the Neanderthals' final demise, which occurred around forty-two thousand years ago. However, in the millennia that followed the invention of the bow around fifteen thousand years ago, up until the end of the Neolithic, there seemed to be little further innovation in weaponry.

It was the advent of metalworking in the Copper and Bronze Ages that resulted in two important revolutions in weapons technology that transformed our ability to kill each other. The first invention, the weapon most celebrated by military enthusiasts of all ages, was the sword, the world's first purpose-built antipersonnel weapon. Swords are basically flattened clubs with sharp edges, and they are wielded in much the same way in battle: soldiers swing them about using a three-stage sling action. But they have two main advantages. Because they are lighter and don't have a heavy head, they can be swung about more rapidly, so a swordsman can easily defeat someone wielding a clumsy club or ax. Being lighter, swords do not build up so much kinetic energy, but since a sword stroke has a slashing action, like that of a carving knife, it can cut more efficiently through flesh and cause traumatic muscle injury and blood loss. And once the swordsman has incapacitated their enemy with a cut, they can speedily finish them off by using the point of the sword like a spear in true Hollywood fashion. In the Bronze Age, swords quickly became the weapon of choice for elite warriors, the earliest recorded war using these weapons occurring in Mesopotamia about 2700 BCE when Enmebaragasi, king of the Kish, led the Sumerians to a victory over the Elamites. It was the first of what would become an almost constant series of conflicts in the region, as the Bronze Age city-states grew in size and ambition.

Though infantry swordsmen make excellent fighters, they are slow moving and cannot sustain a high continuous power output. The second key military advance in the Bronze Age, therefore, following the invention of wheels, was the introduction of the chariot. The Sumerians built battle chariots as long ago as 2000 BCE, harnessing asses to two-wheeled carts fitted with solid wooden wheels. These clumsy, slow-moving vehicles lacked any suspension and would have been uncomfortable and hazardous to ride in. However, chariots were steadily improved over the next thousand years by using flexible spoked wheels and by replacing asses with horses. This made them much faster, more comfortable, and more formidable vehicles. Homer's *Iliad* suggests that the Mycenaean Greeks and Trojans were the masters of chariot warfare, and that they used their vehicles in spectacular single combats. But in real life, generals used squadrons of chariots as shock forces to carry elite troops to the heart of the battle and help break through enemy lines, just like later cavalry regiments. The biggest chariot conflict in history was probably the Battle of Kadesh in 1274 BCE, between the forces of the Egyptian Pharaoh Ramesses II and those of the Hittite King Muwatalli II. Two thousand chariots formed the spearhead of each army, and both sets caused havoc among enemy infantry, but despite mass casualties on both sides, the result of the battle was inconclusive.

Their weaponry may have looked impressive, but the armies of the Bronze Age were small by later standards, especially considering the size of the states they defended. At Kadesh, each side fielded only around fifteen thousand soldiers, a tiny number compared with the hundreds of thousands that the Persians put out to fight Alexander the Great, a thousand years later. The biggest disadvantage of Bronze Age weaponry was just how expensive it was. Ramesses II built a huge military complex at his capital Pi-Ramesses in the Nile delta, complete with stables and a bronze foundry that made spearheads, swords, and the tackle for his chariots. But only the largest, wealthiest states could afford to arm themselves in this way, and even Ramesses's army was small, so during the late Bronze Age the Middle East became dominated by a few large empires: the Egyptians; the Hittites in Anatolia; the Assyrians and Babylonians in Mesopotamia; and the Mycenaeans in Greece. Despite continual warfare among themselves they proved invulnerable to attack from outside because smaller, poorer states simply could not afford to equip an army that could challenge them. All this, however, was to change with the advent of iron.

The first shock was the coming of the Sea Peoples, who in about 1200 BCE mysteriously appeared and attacked Egypt and the other states of the eastern Mediterranean. Their appearance coincided with the collapse of all the major Bronze Age empires in the region, but archeologists are divided about whether the Sea Peoples caused the decline and where they actually came from. There is some evidence that they originated in the northern Mediterranean: from the Aegean, northern Italy, and Sardinia. Bas-reliefs on the Temple of Ramesses III at Medinet Habu depict them sporting horned helmets and holding long swords and round shields, like modern-day pictures of Vikings. Some archeologists have suggested that they were exploiting improvements that their smiths had made in their bronze work, and that they later developed swords made out of iron to produce larger, more lethal weapons. It seems likely that the advent of iron drove a second technological revolution in weaponry and finally put paid to the dominance of the huge Bronze Age empires.

Iron had several advantages over bronze for making and supplying weapons. Iron is around 15 percent lighter than bronze, so it could be made into longer swords that were still light enough to use. Smiths could also fine-tune its properties. As they hammered swords into shape, they could repeatedly fold the metal over, making the main body of the sword tougher and more resistant to breaking, while they could temper the edge to harden it. Iron and steel swords would have been better weapons than ones made out of bronze, and iron woodworking tools would have been harder wearing and sharper, enabling ship-wrights to build better sailing craft to transport troops. But the biggest advantages of iron were its ready availability and cheapness. Any village blacksmith could churn out weapons and tools, so small states with little infrastructure could still equip an army of infantrymen and build a navy that could challenge the largest.

As people bred larger and larger horses, small states could also supply their own cavalry much more cheaply and effectively, since warriors could simply ride a horse, rather than the state having to build a chariot and equip it with a pair of ponies. By the time of Alexander the Great, cavalry had long superseded chariots, and Alexander demonstrated its superiority in his victory over the Persians at the Battle of Gaugamela in 331 BCE. Though the Persians smoothed the battleground the night before the battle to make it easier for their chariots to roll across, their chariot charges proved ineffective against the well-trained Macedonian troops. They simply opened their ranks to let the chariots pass harmlessly through, before the infantry dispatched the charioteers from behind. Meanwhile, Alexander, on his faithful steed Bucephalus, and his crack cavalry squadron charged into the Persians' left flank and routed them. Cavalry, which were faster and more maneuverable, and which allowed the riders to swing down their spears or curved sabers on the enemy, were to remain the strike force of armies for the next two millennia.

The overarching effect of these far-reaching military developments was that wealthy states became far more vulnerable to raids and larger attacks from outside their borders, while ambitious rulers of smaller states could concentrate on building up a large military force to overcome existing empires and enrich themselves. Time and again, wealthy empires became enticing targets for "barbarians," and empires rose and fell at an ever-increasing rate. In the Fertile Crescent, the established powers of Assyria and Babylon were overcome first by the Medes, and then by the Persians, led by Cyrus the Great. But the Persian Empire itself was not destined to last long, as less than two hundred years after its birth, it fell to Alexander the Great, who hailed from the obscure northern-Greek province of Macedonia. The highly productive Nile valley was subjected to an even more rapid series of invasions, as Egypt was conquered in turn from the east by the Assyrians; from the west by the Libyans; from the south by the Sudanese; from the east again by the Persians; from the north by Alexander the Great and his general Ptolemy; and finally from the northwest by the Romans. In East Asia, the wealthy rice-growing areas of southern China were fought over for several hundred years by a number of players in the Warring States period, before being held temporarily by a series of dynasties: the Chin, Han, Jin, Tang, and Song.

But no ruler is happy to give up their empire, so over the two millennia that followed the start of the Iron Age, states invested heavily in their military. The most successful states bankrolled engineers to develop new forms of weaponry that would give them the military advantage that allowed them to expand and keep their neighbors at bay. The best known of the engineers was the Greek mathematician Archimedes, who came up with a variety of weapons to repel the Roman siege of Syracuse in 213 BCE. He devised improved catapults; a crane to lift Roman ships out of the water; and other stone-throwing devices. There is no evidence, however, to support the story that he used mirrors to focus the light of the sun and set fire to Roman ships.

But it was the Romans who took military engineering most seriously and who had whole sections of its army devoted to it. Corps of workers known as *fabri* served in each legion, while sections devoted to the construction of military camps were led by officers known as *immunes*. The *immunes* also organized the construction of bridges, walls, aqueducts, and most crucially the fine straight roads that enabled Roman armies to march rapidly to and from trouble spots carrying their baggage train with them. Military engineers such as the writer Vitruvius also developed a wide range of powerful weapons for use in battles and sieges.

The first was the ballista, invented by Greek engineers working for the early fourth century BCE Syracusan dictator, Dionysus I, but later perfected by the Romans. The ballista was basically a multi-man spear-throwing machine with an arm projecting out on each side like the arms of a crossbow, and with their tips linked by a rope, like the string of a bow. The difference was that the arms were rigid and were not joined together in the middle, but mounted on hinges, around which were wrapped spiral tendon springs, The operators drew back the rope using a winch and ratchet mechanism, pulling the arms back and storing energy in the tendons, before loading the bolt and finally freeing the ratchet to let loose some five hundred joules of energy, propelling the bolt up to four hundred yards (365 meters).

A Roman soldier operating a ballista. Note the twisted tendons that power the machine and the winch to draw it. This model seems to have been modified to fire a stone rather than a bolt.

The Romans used the same elastic storage mechanism to power another siege engine, the onager, or ass. Instead of having two arms, this machine just had a single long arm with a basket at the end, which rotated around a horizontal axis. The operators pulled back the arm until it was horizontal, like the bar of a mousetrap, and placed a rock in the basket before releasing the catch to hurl the rock against the enemy. The downside of the early onagers was that much of the energy was wasted accelerating the arm, which could also damage the machine when the arm was stopped at the end of the throw. Later on, therefore, engineers added another element to the arm, a rope at its end in which to hold the rock. As the arm rotated forward, it accelerated the rock at the end of the rope, like a slingman throwing a stone. Later onagers could throw rocks farther and faster and were more efficient and safer than the early ones, since the arm was decelerated toward the end of the throw and came to a gentler halt. These onagers could throw a hundred-pound (45-kilogram) rock up to four hundred yards (365 meters), giving the rock around one hundred kilojoules of kinetic energy. They were over a thousand times more powerful than a single human being.

The engineers of southern China were just as resourceful and inventive as the Romans. The Chin dynasty, for instance, was the first to complete a huge defensive wall around their realm. They linked up a series of walls constructed in the seventh century BCE to create the Great Wall of China in around 220 BCE, protecting the dynasty from raiders in the north. The Chinese also developed a series of effective weapons that were later taken up farther west.

The first was the crossbow, invented in the fifth century BCE, which used the same principles as the wooden longbow, but which had stiffer bamboo or metal arms. Crossbows were fitted with a winch and ratchet arrangement so that the user could pull back on the string with greater force and store more energy in the arms. They could therefore fire their bolts at a higher speed than with longbows, so they had a longer range and packed a bigger punch; bolts were given up to 130 joules of kinetic energy, more than enough to penetrate the strongest armor and go right through a human body. Their main disadvantage was their slower rate of fire; users could only fire two bolts per minute. But the Chinese overcame this problem with their invention in the fourth century BCE of the repeating crossbow, the bolt-action rifle of the day. After each firing all the user needed to do was to pull back with one hand on a lever that simultaneously drew the string back and allowed a new bolt to pop up from a magazine. Repeating crossbows could fire as many as ten bolts in twenty seconds, but they were less powerful that the single-action devices and far less accurate, since they had to be fired from the hip. They were used mainly for defense and, because they were easy to load, were even recommended for the use of ladies!

In the fifth century BCE the Chinese also developed much larger weapons—trebuchets—which used a sling action like the later Roman onagers. The first traction trebuchets were essentially simple wooden beams, hinged a quarter of the way from one end, with a sling mounted at the end of the long arm and handles at the end of the short arm. The operators loaded a rock in a basket at the end of the sling, then all together pulled down on the handles to swing the long end up and accelerate the rock forward and upward at the enemy. Trebuchets were imported into the Middle East from the fifth century CE, where they were used as siege weapons by the triumphant Islamic armies. They were finally adopted in Europe and equipped with a counterweight, instead of being powered directly by muscular action. The operators loaded them using a crank or treadwheel to ratchet up the device, just as the Romans had done with their onagers, pulling down on the long arm and raising the counterweight, making the weapon even more powerful and accurate.

Trebuchets reached the peak of their development in medieval times, probably the most famous example being Edward I's massive machine,

Warwolf. This terrifying device was over sixty feet (18 meters) high and was capable of throwing a three-hundred-pound (140-kilogram) rock several hundred yards, projecting some five hundred kilojoules of energy, a thousand times the power of Roman ballistas and five thousand times as powerful as a single person. Edward used it just once, in 1304 during his wars of conquest against the Scots, to attack Stirling Castle, the last stronghold of the Scottish armies. The defenders were so petrified by the prospect of the attack that they sued for peace before Warwolf could be used, but Edward refused to let them surrender until he had tested the machine; in four days it totally destroyed one of the castle's curtain walls.

Despite their ingenuity, the efforts of the Romans and Chinese to protect their empires using novel mechanical devices was only partially successful. One problem was that they still required muscular effort to load them, so they had to be manned by large numbers of people, and they were so large that they were hard to transport and put together. The individual components of Warwolf, for instance, filled thirty wagons, and it took several days to assemble. The Romans certainly managed to use their engineering expertise to reverse the usual historical trend and expand their empire northward into barbarian territory. But Roman armies were always ponderous and vulnerable to ambushes at the empire's borders; whole legions fell foul of Germanic and British tribes.

The Romans also had little answer to the tactics of a state on their eastern border, the Parthians. A people who ruled over Asia Minor (now Iran) from 250 BCE to 220 CE, the Parthians were expert horsemen and specialized in mounted archers. The key to their success was their use of the composite bow. Combining a back made of bone and a front made of sinew, composite bows are much better at storing energy than wooden longbows and so could be made much shorter and be used on horseback. The Parthian tactic was for their mounted archers to approach an enemy line, and, when it surged forward, to retreat and to fire behind to cause panic. A Parthian army destroyed the far more numerous army led by the Roman billionaire Marcus Lucinius Crassus in 53 BCE.

Eventually, even the might of Rome crumbled in the fifth century CE. The western empire was infiltrated by Vandals and Visigoths from the north, while the eastern empire was later attacked from the east by Islamic armies that had originated in the deserts of Arabia. From the seventh century onward the Arabs overran Persia, Turkey, and Egypt, even reaching as far west as Tunisia, Morocco, and Spain. The relatively poorer northern territories of the former Roman Empire, Britain and Gaul, were also plundered and eventually conquered by people from even farther north, the Angles, Saxons and Jutes, from northern Germany and Holland. And even these invaders themselves became the victims of people from farther north still, the Vikings of Denmark and Norway, who terrorized the kingdoms of the British Isles and northern France.

Before 800 CE, Chinese engineers had been just as unsuccessful as their Western counterparts in using their technology to maintain stable regimes. They had crossbows and trebuchets, and in the fifth century CE they invented stirrups, which enabled Chinese cavalrymen to stabilize themselves on their horses and wield their swords and spears far more effectively. But these inventions soon became common property and could be used against them. Instead, the technological game changer in warfare was the invention in ninth century China of gunpowder. The possession of this new secret weapon, which was more powerful than any that preceded it, gave the Song dynasty a decisive military advantage.

It all happened by accident, however. In their search for the elixir of life, Taoist alchemists paradoxically discovered what would become the harbinger of death. Composed of a ground mixture of saltpeter (potassium nitrate) and sulfur, along with some charcoal to provide the initial combustion, gunpowder burned rapidly, even in the absence of oxygen, to produce large volumes of hot gas in the reaction:

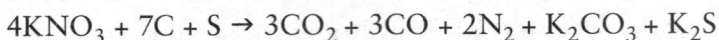

$$4KNO_3 + 7C + S \rightarrow 3CO_2 + 3CO + 2N_2 + K_2CO_3 + K_2S$$

The reaction is not only rapid but powerful, releasing a large amount of energy, some three megajoules per kilogram, which can cause the gases it produces to expand and create flames. The early mixtures the alchemists produced contained around 60 percent saltpeter and burned only slowly, so the Chinese used it mainly as a means of setting fire to enemy positions—adding it to the tips of their arrows, or coating it around the projectiles thrown by their trebuchets.

Increasing the amount of saltpeter to 70 percent speeded up the reaction, producing what was more like an explosion, and gradually Chinese military engineers learned how to harness the energy released to produce virtually all of the weapons that we have come to think of as Western inventions. The engineers packed relatively slow-burning gunpowder into open-ended tubes to produce flamethrowers to point at the enemy; or they reversed the tube and added a stick to its rear to produce rockets. They gradually learned how to use even the most dangerous, fastest-burning gunpowder. They packed it into thin metal casings along with stones and metal shot to produce grenades, mortars, and bombs, which they flung or fired at enemy positions, where the weapons exploded with devastating effect. As China's metallurgy improved, the engineers packed gunpowder into the base of cast-iron tubes to enable them to fire projectiles directly at the enemy. In the earliest examples, they poured multiple pieces of stone or metal into the tube to produce eruptors or shotguns; later on they used single balls or stones that fitted more tightly into the barrel, to produce the world's first cannons and handguns

Gunpowder stores so much energy that just one gram is enough to instantaneously produce three thousand joules of energy, thirty times the energy that a person could inflict and theoretically enough to kill tens of people. This made gunpowder an ideal source of portable fire-power for the Song's armies. Their soldiers no longer needed to work hard to kill their enemies; they merely had to point their weapons and fire. Their armies could be smaller, travel lighter, faster, and still be more formidable. The Song protected the secret of gunpowder weapons for three hundred years. It was one of the main factors that enabled them to keep in power for centuries, repelling all attacks along the Great Wall. The dynasty only fell after the secret of gunpowder finally leaked out of China, in about 1200 CE, and was enthusiastically taken up by its Mongolian enemies in the north. Under the generalship of Kublai Khan, the Mongols were also able to deploy their crack mounted archers as the Parthians had before them, to finally defeat the Song in 1279 CE.

Gunpowder weapons transformed the history of Eurasia even more dramatically than the advent of bronze and iron weaponry. Its fundamental effect was to restore to large, wealthy states the military advantage over their smaller neighbors that they had lost with the advent of iron. Gunpowder was always difficult and expensive to produce, largely because it was so hard to obtain its main ingredient, saltpeter. This forms naturally on warm, damp surfaces, but only in small quantities, and it had to be scraped from the earth or from the walls and floors of agricultural buildings and be purified by a long process of solution and crystallization. It could therefore only be obtained in great enough quantities by states that were large enough and well-enough governed to organize a workforce to collect the saltpeter and to set up and run gunpowder factories.

In China, where the secret of gunpowder had sustained the Song, and after the short, brutal rule of the Mongols, gunpowder weapons enabled a new dynasty from southern China, the Ming, to rule for a further three hundred years. Southern and western Asia, meanwhile, became dominated by two vast Islamic empires, both of which were innovative in their use of gunpowder technology and which emerged in the fifteenth and sixteenth centuries. The Ottomans, originally from a small principality in Anatolia, were particularly skillful in making and using cannon. Mehmed II employed them successfully, for example, in the final siege of Constantinople in 1453, and the Ottoman army also developed a way of using cannon successfully in battles. They mounted a line of guns on a series of wagons behind their infantry and used them to devastate the enemy lines before they charged. These tactics enabled the Ottomans to defeat the Mamluks in Egypt and take control of an empire that stretched from Hungary and Serbia in the west to Azerbaijan in the east. They dominated the Middle East well into the nineteenth century. In South Asia, the Moghuls, who originally hailed from present-day Afghanistan, used similar tactics to build an empire all the way across modern-day India, and to maintain it until finally being challenged by the French and British colonial armies in the late eighteenth century.

Europe was late to obtain gunpowder, but embraced its use so enthusiastically that the continent quickly started to make a massive impact on world history. The first weapons they developed were cannon, which superseded trebuchets as siege weapons. They quickly rendered conventional castles obsolete, since the stone and iron cannonballs they fired had enough kinetic energy to break through even thick stone walls. James II of Scotland's huge seven-ton cannon of 1449, Mons Meg, for instance, could fire 380-pound (175-kilogram) cannonballs at speeds of 675 mph (300 meters per second), giving it a range of two miles (3.2 kilometers). Each ball contained around eight megajoules of kinetic energy, making the cannon around fifteen times as powerful as Edward I's Warwolf trebuchet and eighty thousand times as powerful as a single person. An arms race quickly started, as defenders built shorter fortresses with increasingly thick walls and projecting bastions to deflect the cannonballs, while besieging armies built ever-larger and more powerful cannon.

European gunsmiths also developed handguns, such as harquebuses and muskets, to be used on the battlefield. Even these small early weapons were quite powerful. A sixteenth-century musket, for instance, could typically fire a 1.4-ounce (40-gram) bullet at speeds up to 120 mph (50 meters per second), giving it kinetic energy of five kilojoules, easily enough to kill a man; the shock waves from the bullet would spread outward as it passed through the human body, leaving a cone of destruction and a gaping exit wound. Muskets could have a lethal range of over one hundred yards (90 meters).

The new weaponry certainly made battles in Europe more bloody, but none of the states were ever able to develop a monopoly on the new military technology, so Europe continued to be riven by interminable and inconclusive wars between the large numbers of small competing powers. Gradually, however, the difficulty in producing adequate amounts of gunpowder led to the inevitable demise of many of the smaller European cities and dukedoms. Europe began to consolidate into larger states and kingdoms, leading to the formation of some of the western European countries of today: France, Portugal, and Spain.

The determination of these maritime states to muscle in on the trade with China and India that had traditionally flowed inland along the Silk Road led them to voyage around Africa to the East. Their need to defend their merchant ships led them to establish navies armed with formidable cannons. And the mistaken belief of Christopher Columbus that sailing westward across the Atlantic was a quicker way to China led him to discover the New World. Once there, the greater firepower of European soldiers, armed with iron swords, cannons, and guns, and mounted on fast-moving horses, gave them a decisive military advantage. These factors enabled a few hundred adventurers to defeat and take over the indigenous empires of the Aztecs and Incas, giving birth to the age of European colonialism.

Chapter 13

RAISING THE POWER SUPPLY

According to the history that most of us are taught in school, the watershed between the primitive medieval age and the modern world was the Italian Renaissance. The rediscovery of the ancient learning of classical Greece and Rome led to a flowering of the arts and sciences in the city-states of Florence, Rome, and Venice that transformed European thought, swept away medieval superstition and ignorance, and led inexorably to industrialization. The story certainly seems to ring true at first glance. After all, who is not amazed by the quality of Venetian glass, the mastery of perspective in the paintings of Donatello and Raphael, the lifelike flesh of Michelangelo's sculptures, the engineering ingenuity of Leonardo's sketchbooks, the bravado of Brunelleschi's dome, or the perfect proportions of Palladian villas? Even today the achievements of the fourteenth- and fifteenth-century Italians dazzle the amazed tourist.

Yet beneath the cultural veneer, little had in reality changed in Italian life. The economies of southern Europe had not been transformed, merely temporarily enriched by trade. Venice had prospered because it lay at the end of the Spice Roads and acted as a market for goods imported from the East; Florence became Italy's banking hub; and from the sixteenth century onward, Rome, along with the cities of Spain and Portugal, decorated its churches with gold and silver plundered from

153

the colonies of the newly discovered Americas. Local artists and artisans may have benefited from the patronage of a few wealthy families, but money continued to flow away to Asia, and there was no net gain in the prosperity of the general population. By the end of the fifteenth century, the Florentine republic had collapsed and the general population were happy to follow the exhortations of the ascetic Dominican friar Savonarola and consign artworks to the flames in his notorious Bonfires of the Vanities. In 1600, over two hundred years after the start of the Renaissance, 86 percent of the people of Italy still worked in agriculture, and the country did not industrialize until late in the nineteenth century. Which is why so many impoverished Italian peasants emigrated to seek a better life in Britain and America until well into the twentieth century.

In truth, the only really sustainable way in which a society can improve its standard of living is to increase its material productivity. It has to find a way to produce more food to eat; to find a way to produce more useful mechanical power; and to find a way to produce more useful heat energy—all the while minimizing the work that people have to expend themselves. As we shall see, the changes that really transformed the economy of Europe took place far away from Italy, and they were effected by people who had received none of the benefits of a classical education.

During the Renaissance, Europe was in fact an economic and political backwater. The wealthiest country in Eurasia was China, which prospered because of the much-higher productivity of its agriculture, especially in the rice-growing south. Like the farmers of the Middle East and central America, those of southern China were masters of irrigation. They grew rice in small paddy fields, gradually raising the water level to support the plants as they grew. They used lightweight wood and bamboo water ladders, like saqiyahs but powered by people, which they moved from field to field. And in the fourteenth century they invented vertical-axis windmills that could pump the water for free. Rice growing was labor-intensive, like most forms of cereal farming, but the Chinese developed numerous techniques to increase its yield. They raised the soil fertility in several ways. They exploited the tiny floating fern *Azolla*, which has a symbiotic relationship with nitrogen-fixing cyanobacteria, allowing it to grow in their paddy fields before plowing it in. And they manured their crops with both animal feces and human night soil. They also interspersed rice harvests with catch crops of soybeans and wheat to reduce the buildup of disease. Managed in this way, their fields could produce two or even three harvests a year, and southern China acted as the rice bowl of the country, producing vast quantities of food, as much as two to three tons per acre (5–8 tons per hectare) per year, much of which was transported up the Grand Canal or by sea to the big cities of the north. Farmers were also able to grow vegetables and mulberry trees on the bunds between the paddy fields, allowing them to be self-sufficient in food, while they used the mulberry leaves to farm silk moths. They harvested the cocoons and processed the silk in between planting and harvest seasons, using a range of ingenious spinning and roving machines.

The whole system, a hybrid of agriculture, horticulture, animal husbandry, and light industry, created a surplus that supported a stable government, and a professional bureaucracy. The state passed on information about best practice using mass-produced woodcuts that could be read by the literate population. And Chinese engineers were also just as mechanically inventive as the Romans'. They developed a range of water mills to grind flour and power other small-scale industries, though most of their waterwheels were inefficient vertical-axis devices. However, this mostly deforested country had only a limited supply of wood to burn, which prevented the development of large-scale industries and limited overall productivity. Industries were either small-scale or localized in the few areas where forests remained. Porcelain production, for instance, was centered in the heavily forested mountain region around Jingdezhen, nine hundred miles from Beijing.

Finished pieces had to be carried by hand over mountain passes and across lakes, incurring high transport costs that meant that porcelain was always a luxury item.

The situation was quite different in Europe, whose farmers continued to practice the labor-intensive but unproductive large-scale cultivation of rain-fed cereals, growing monocultures of wheat, barley, rye, and oats. The bulk of the land was worked by uneducated peasants, whose methods had changed little since the early days of agriculture; they still broadcast their seed and weeded the crop by hand, harvested it with sickles, and threshed it using flails. The main technological development, which had been imported from China, was the use of the moldboard plow, which not only scratched the soil surface like the primitive ard, but turned over the sod, burying and suppressing weeds. This reduced weeding effort and allowed farmers to exploit heavier, richer soils, but the flat moldboard was hard to draw through the soil. It had to be pulled by a team of four or even eight oxen, so farmers had to set aside more land for pasture or to grow oats to feed their beasts. Laborers also had to spend much of the late winter breaking up clods of earth with mallets. Without any appreciation of the techniques the farmers could use to increase soil fertility, crop yields were low, typically around a ton per acre (2.5 tons per hectare), and land could not continually be cultivated. Typically after a crop of cereal, farmers grew legumes such as clover the next year, before leaving the land fallow for a further year. Consequently only two-thirds of the land was under cultivation at any one time. The overall productivity of European cereal farmers was consequently a fraction that of China's rice growers, thus supporting few artisans and only a tiny ruling elite.

Two countries on the North Atlantic coast, the Netherlands and England, also had the additional problem that, because of their high population density, they had cleared most of their land for agriculture and had, like China, reduced their forest cover to below 10 percent. This limited their supply of their main source of heat energy—firewood.

Paul Warde of the University of East Anglia estimated that in sixteenth-century England the energy people obtained from wood was limited to twenty terajoules per year, only slightly more than the energy they obtained from food. Paradoxically, though, it was these two countries that finally broke through the barriers that had previously limited economic activity in Europe. The key to their success was a newfound ability to exploit synergies between innovative forms of food production, renewable sources of mechanical power, and the pioneering use of fossil fuels. Time and again, advances in one area helped spur on the others, generating a virtuous circle that led to wealth and plenty.

The Netherlands led the way. Lying on the Atlantic delta of the Rhine, it was well-placed for trade, not only with its traditional partners in the Baltic, but also with the New World, Africa, and the East. Over the fifteenth and sixteenth centuries the country built up a fleet of merchant ships and a navy to protect them. It imported timber from the Baltic, wool from England, and spices from the East Indies. However, the land area of the Netherlands was small, and much of its western region was low-lying, prone to flooding from the North Sea and unsuitable for cereal production. These problems led the merchants who dominated the politics of the region to innovate. They developed a major herring fishery, and together with the wealth they obtained from trading, this allowed them to pay for grain, which they imported from the Baltic. Their farmers also developed novel techniques to manage their limited land area to produce the other foods that they needed in a novel form of highly productive land management: market gardening.

Over the centuries before, Dutch traders and their Spanish rulers had started to introduce new high-yielding root crops to the country, crops that the Moors had brought into southern Europe from farther east, most notably turnips, parsnips, and carrots. Dutch farmers combined growing these new crops with traditional ones such as onions, beans, vetches, and hops and put out the remainder of their land to grass to feed cows for dairying. The farmers developed increasingly complex crop rotations that kept a constant ground cover and reduced the buildup of diseases, and they spread the manure from the cows over their land to maintain soil fertility. The new techniques produced a sophisticated form of horticulture that was labor-intensive but highly productive. This provided the ever-expanding population with a rich diet and provided the foundations of Holland's modern-day horticulture industry.

But the Dutch also sought to grow more food by reclaiming land from the sea and draining low-lying bogs and fens. The fifteenth century saw the start of a particularly ambitious project to exploit the peat bogs of the huge Holland region, an area of some 250 square miles (650 square kilometers) that lay at or below sea level. The project was to a large extent driven by the need to obtain the fuel that the deforested country desperately needed. They cut the peat and dried it over the summer months before ferrying it to the towns and cities around the perimeter of the region—Naarden, Utrecht, Gouda, Rotterdam, Delft, Leiden, Haarlem, Alkmaar, and Amsterdam—along the lakes and canals that the flooded peat cutting automatically opened up. Peat, which is basically fossilized moss, is a poor fuel, containing only a tenth of the energy of wood, but the quantities of peat the Dutch obtained were huge. Jan de Zeeuw of the Agricultural University of Wageningen estimated that the cuttings produced twenty-five petajoules of energy each year, over three times the energy per head of population that the woodlands of England were producing at the time. The efficient water transport allowed the supply of peat to the very center of the towns, to power a whole range of new urban industries. Lowland peat fueled the Dutch Golden Age of 1600 to 1700; it heated the clamps that made the bricks, pantiles, and lime mortar that built the fine cities; fired the kilns that produced the fine Delft pottery; and heated the breweries and dye works. And once the peat was cleared, the Dutch drained the land.

The first task was to build a dike around an area of land, and the next to pump the water out to sea. Since the land was so flat, the

Dutch could not use waterwheels to provide the power to drive water-lifting devices, so they turned to the one renewable-energy resource that this flat, exposed region had in abundance—wind. The Islamic states around the Mediterranean had been using windmills for several centuries to power pumps to irrigate their crops; they were common throughout Spain, hence their appearance in *Don Quixote*. However, these were small, inefficient machines powered by simple cloth sails. Dutch engineers, most notably the polymath and inventor of decimals, Simon Stevin, transformed them into complex multistory buildings with a rotating upper level that the miller could turn into the wind, and huge wooden sails, twisted like modern day propellers, and fitted with slats that acted like the separated primary feathers at the wing tips of eagles. They used these powerful machines—a typical mill could generate fifteen to twenty horsepower (11–15 kilowatts)—to drive scoop wheels to raise water from a lower basin up into a canal from where it could flow to the sea. Finally, the Dutch could exploit the rich clay soils that had been exposed by the peat diggings and produce yet more food. They had demonstrated for the first time that people's ability to harness power can actually enable us to alter the face of the earth.

Together, the three energy sources—market gardening, windmills, and peat—made the Netherlands the wealthiest country in Europe between 1600 and 1700. For the first time, people of all social classes could eat a varied diet, afford fine furniture and crockery, and even buy paintings to adorn the walls of their homes. Dutch industry financed a much more broadly based artistic flowering than the Italian Renaissance. And Dutch art did not just illustrate religious stories, but showed all aspects of life: from intimate domestic dramas; riotous alehouse and brothel scenes; the public life of markets, churches, and streets; to the labors of workmen in the fields and at sea. Painters such as Rembrandt and Franz Hals could depict character with profound psychological insight, and Vermeer could depict light with unprecedented brilliance.

IIIIIIIIIIIIIIIIIIIII

The emergence of England started later than that of the Netherlands, but the country had several geographical advantages over its neighbor across the North Sea that enabled it to harness much more power and use it to generate much more sustained economic growth. The first advantage, often regarded by its inhabitants as one of the banes of English life, was its mild, damp climate. In mediaeval times, England was less densely populated than the Netherlands, but the long growing season allowed it to be self-sufficient in cereals, even though its villagers cultivated the land in a form of unproductive collective farming. Sheep and cattle also thrived in the mild climate, feeding on the grass that grew well in the moist soils. This enabled large landowners such as the Cistercian monasteries to set up a thriving woolen industry, which expanded further after the Black Death had killed a third of England's population. After the pandemic, the lack of labor meant it was more profitable to convert farmland into huge sheep pastures, particularly in the light soils in the east. The farmers made large profits from exporting wool, profits that built the famous churches of the region, while the sheep manure started to improve the exhausted soils.

English farmers were also well placed to benefit from the example set by the Netherlands. In East Anglia, economic growth was fueled, as in Holland, by peat diggings, most notably in the fens around the Norfolk coast, where peat was cut and shipped upriver to England's premier wool center and second-largest city, Norwich. However, compared to the Dutch peat industry, the work was amateurish and small-scale, and the peat cuttings were never drained but left as flooded lakes: the Norfolk Broads. The industry was soon abandoned and forgotten, and until the middle of the twentieth century the broads were simply regarded as a natural landscape feature and used as a playground of the rich for their sailing holidays.

During the political and social turmoil of the English Civil War in the middle of the seventeenth century, a newly emerging class of yeomen farmers imported agricultural ideas and expertise from the Netherlands and promoted a revolution in English agriculture and industry. Landowners brought in Dutch engineers to help drain the wetlands of eastern England, most notably the Earl of Bedford, who employed Cornelis Vermuyden to drain the Cambridgeshire fens, creating five hundred square miles (1,300 square kilometers) of rich, productive farmland. Smaller farmers also started to consolidate land ownership, converting land that had previously been split into scattered strips and collectively farmed in the Champion system into individual farms, a system known as Several. This enabled them to try out new techniques, notably those promoted by Dutch immigrants such as Samuel Hartlib. Gentlemen farmers were particularly innovative and eventually devised a typically English compromise between extensive cereal growing and intensive market gardening. They started to intersperse years growing wheat with years growing root vegetables, legumes, and other catch crops. This eventually led to the archetypal, Norfolk four-course crop rotation, in which a year growing winter wheat was followed by ones growing in turn clover, turnips, and spring barley, allowing the land to be cropped continuously.

Individual farms could also combine arable farming with the husbandry of sheep and cattle. Farmers pastured their beasts on grass leys and allowed them to feed in the turnip fields, a technique that had the additional benefit of providing manure to fertilize the fields. The result was a mixed farming regime that more than doubled the productivity of the land and enabled England's farmers to feed a population that rose from 3.4 million in 1560 to 6.6 million in 1750. Around the expanding cities, smallholders imported Dutch horticultural techniques to feed the growing population. The London suburb of Chelsea was a particularly busy center of market gardening; the growers fertilized the fields with the night soil from the city and produced high yields of fruit and root vegetables. Chelsea has now long since been swallowed up by the London conurbation, but the legacy of the industry remains in the guise of the Chelsea Flower Show, the greatest gardening show in the world, and Chelsea Physic Garden, which on weekends has probably the highest concentration of posh men in red trousers on the planet.

England's clouded hills gave it a further advantage, since the year-round rainfall fed flowing streams and rivers that millers could reliably harness to provide almost continuous mechanical power. Landowners could build water mills, which were expensive to construct but far cheaper to run than windmills, and they could site them every few hundred yards down England's rivers. The Domesday Book of 1088 shows that even in Saxon times England had over six thousand water mills; by the end of the fourteenth century there were over ten thousand, the Cistercian monks being particularly adept engineers. Many of the mills were used for the traditional purpose of grinding grain, but millwrights also adapted them to power other heavy industries.

One of the few downsides of water mills and windmills is that they produced a rotational motion. This is fine for some processes, such as grinding grain and scooping up water, but most of the heavy industry of the ancient and premodern world demanded a quite different motion: reciprocating. The machinery needed to mimic the back-and-forth actions of workpeople and their traditional tools. In the textile industry, for instance, woolen cloth needed to be hammered with mallets to loosen the nap and cause the threads to entangle more closely with one another, shrinking and thickening the cloth and making it more waterproof, a process known as fulling. Ironmasters needed to ventilate their furnaces with bellows to smelt their iron, and they needed to hammer the soft bloom their furnaces produced to remove slag and convert it into bar iron. Sawyers needed to swing their saws back and forth to cut tree trunks into planks. To mechanize these processes early engineers needed to devise techniques to convert rotational motion to reciprocating motion.

The first solution was probably the cam, which was simply a peg mounted onto an extension of the axle of a wheel, called the camshaft. In fulling mills and forges, these pegs made intermittent contact with extensions to the handles of huge trip-hammers, lifting them around their hinge and then releasing them, so that they fell under their own weight onto the cloth or iron bloom. The second solution was to reverse the operation of the crank that people had been using to power their querns. One way of doing this was to link the crank to a rod known as a slider, which was constrained to move only along it length. Sliders were incorporated in sawmills from Roman times onwards. In the third century CE Roman sawmill at Hierapolis, Turkey, for instance, the wheel drove two saws back and forth. Another method to produce linear motion, which was used mainly to pump water, was to reverse the action of the treadle that people had been using to power their spinning wheels. Instead of driving a slider back and forth, the crank drove a hinged beam, known as a rocker, up and down, indirectly moving a piston attached at the other end of the rocker. Rockers were first developed by German mining engineers to drive reciprocating piston pumps to drain the silver mines of the Harz Mountains.

None of the mills would be regarded as powerful to modern eyes. Typically, they delivered a mere two to three horsepower, around fifteen hundred to two thousand watts. Nonetheless, this would have been enough to replace the labor of thirty to forty people. And they were also capable of delivering much more powerful blows. Using cams, for instance, a typical iron mill could raise a 220-pound (100 kilogram) trip-hammer to a height of over two yards every second, delivering two thousand joules of energy at each stroke, ten times the capability of even the strongest man wielding the heaviest hammer. Alternatively it could raise five smaller, 22-pound (10 kilogram) hammers to deliver five four-hundred-joule strokes per second. The machinery enabled people for the first time to exceed their physical limits and become superhuman. It could therefore be used to make larger metal components, or to make metal objects faster and using less labor. Machinery was replacing muscle power. Even though each mill had a limited power output, around the same as a small modern lawn mower, together they expanded the industrial capabilities of whole countries. In England and Wales, for instance, Paul Warde, of the University of East Anglia, found that by 1550 mills were producing some 550 terajoules of mechanical energy per year, almost half that produced by the 1.5 million men who made up the country's workforce. By 1700, this had almost doubled to 1,000 terajoules.

Even before the advent of steam-powered machinery, engineers had started to transform whole areas into industrial landscapes, though because the machinery was abandoned over the last two centuries, we have come to regard them as natural. In England, for instance, the rivers of those bucolic refuges of today, the Cotswolds and the Pennines, were once dotted with fulling mills that processed the woolen cloth that had been locally spun and woven from the sheep that still roam the hills. Similarly, the streams of the Weald in southeast England, made famous as the scene of Winnie the Pooh's games of Poohsticks, were harnessed to power the bellows and hammers used in Britain's iron industry; the iron-laden rocks of the area were forged with local charcoal, made from the coppiced woodland that clothed the rolling hills. In its heyday the Wealden iron industry produced some twenty-five thousand tons of iron a year. Nowadays, however, little remains of this industrial landscape, colonized as it is by London commuters, bar the overgrown coppices and unusual place-names, such as Furnace Green and Abinger Hammer.

But the island of Great Britain was blessed with an even greater natural advantage over the Dutch in fuel; the soil hid huge deposits of high-grade coal, laid down during the Carboniferous period in a band that stretched from South Wales up through the Midlands and the north of England to the central belt of Scotland. Coal is a far better fuel than both peat and wood. It contains twice the energy per unit mass, and since it is much denser, it contains five times as much energy per unit volume than wood and fifty times more than peat. It is also deposited in far more concentrated seams, so once it is dug out of the ground, it is far cheaper to transport, especially by water. In a further stroke of luck, in the northeast of England, the Durham and Northumberland coalfield was located close to the surface and in a region that was drained by three navigable rivers: the Tyne, Wear, and Tees. Coal here could easily be mined, wheeled downhill along wooden wagon-ways to temporary harbors or staithes on the rivers, and shipped down the coast. Coal powered the explosive growth of England's capital, London, as early as the fifteenth century, providing the heating fuel that allowed its population to quadruple from fifty thousand in 1500 to two hundred thousand in 1600. At the same time the population of Newcastle, at the center of the coal trade, more than doubled, from four thousand in 1400 to ten thousand in 1600.

In the seventeenth century, industrialists also started to replace firewood with coal, which had the double advantage of being much cheaper, and of burning with a hotter flame. The country started to build up energy-intense industries that soon started to rival and out-strip those of the Dutch. England produced ever-larger quantities of high-quality lead glass; built an alum industry along the Yorkshire coast, metalworks around the coalfields of the West Midlands, and potteries around the Staffordshire coalfield. In the east, areas where stone was rare and people had formerly had to construct their dwellings out of wood, coal fired the bricks that allowed them to build fine new houses, in the Great Rebuilding.

Newcastle coal, shipped up the Thames, powered London's booming economy and, following the Great Fire in 1666, provided the energy to build and run a fine new city. The population had almost trebled by 1700 to 575 thousand. The tax on coal even paid for the rebuilding (in stone and brick) of London's many churches and bridges and the construction of Christopher Wren's magnificent new St Paul's Cathedral.

Restoration London became for the first time a cosmopolitan city and a hotbed of ideas, where workmen, merchants, the court, and a new scientific intelligentsia lived cheek by jowl. The city attracted visitors from all around the world to marvel at the new developments and industries, and, in the case of the Russian Czar Peter the Great, to spy on the expanding naval shipyards.

By the end of the seventeenth century coal was supplying some eighty-four thousand terawatts of heat energy per year, four times the energy that England's woodland could supply. Just one heavy industry had not been converted to using coal as a fuel: the iron industry. The productivity of this industry, concentrated as it was in the Weald, south of London, had been limited by the output of the coppice woodland, and by the seventeenth century it was being outcompeted by iron from countries such as Sweden, which had more extensive forests and consequently cheaper charcoal. The obvious answer was to convert smelting furnaces from charcoal to coal, but unfortunately the high levels of sulfur and other impurities in English coal made iron that was too brittle. The breakthrough was achieved in the coalfields of Shropshire in the west of England, where an ironmaster, Abraham Darby, developed a technique to burn off the impurities. He heated the coal in an air-free retort, to produce a purer form of carbon: coke. Using coke had other advantages as well as its cheapness and availability. The coke was stronger than charcoal, so Darby could build larger, taller blast furnaces without the coal being crushed, and the higher working temperature generated by the coke meant that his furnaces could produce molten metal that could be poured into molds rather than having to be beaten into shape: cast iron. Darby made a fortune making cheap cast-iron pots and pans for domestic use to replace more expensive copper ones, and cast-iron stoves and vats for heavy industries such as alum making.

Coal from other British coalfields had higher sulfur content, so it could not so readily be turned into coke, but by 1750 all the difficulties had been overcome, and cast iron was being produced in vast quantities, used not only for pots and pans, but the frameworks for bridges and mills, and to make cheap iron cannon. The output of British iron took off even faster and had by 1790 magnified over fourfold to reach ninety thousand tons a year. As the amount of iron produced and the output of the British economy rose, so, too, did the input of coal for those industries. Coal production was ramped up from six hundred thousand tons a year in 1600, to 3.5 million tons in 1700, and 17 million tons in 1800. It produced some four hundred thousand terajoules of energy per year, forty times that which colliers could have produced from charcoal.

However, as is so often the case, an advance in one industry often stimulates or requires advances in others. For millennia, miners had been breaking open rocks by the simple time-honored method of hitting them with a hammer and chisel, or a pick. As we have seen earlier in the book, a strong workman should be able to use a triple sling action to power their tool up to two hundred joules. However, deep in a coal mine with shallow shafts and narrow coal seams, miners often had to work crouched or even lying down and could not swing their bodies freely. They would be unlikely to develop much more than fifty or one hundred joules at each stroke. Productivity was therefore limited to two hundred tons of coal per miner per year, and coal mining soon became one of the largest employers in the country; the numbers of miners peaked in 1913 at over 1 million. To speed up production, miners introduced a well-tried explosive technology: gunpowder.

By the seventeenth century, gunpowder had been used in warfare for centuries. However, the Chinese had also converted it into a tool that could be useful for civil engineering projects, employing it in 1541 to blast rock where the Grand Canal crossed the Yellow River. French engineers had also used it to build the Canal du Midi, which in 1681 linked up the Atlantic with the French Mediterranean coast. By the end of the seventeenth century, gunpowder also began to be used in coal mines. Miners would drill a hole through the rock and pack gun-powder into it, before setting it alight via a fuse. The pressure wave produced by the explosion splits the rock around it. Since the energy content of gunpowder is around three megajoules per kilogram, a small charge of around ten ounces (300 grams) could produce around one megajoule of energy, ten thousand to twenty thousand times the energy of a miner's pick, and release large quantities of coal, which could be separated from the surrounding rocks and taken to the surface.

As coal production increased, the miners had to dig deeper and deeper beneath the ground, and pits came up against the same prob-lem that had long plagued the silver mines of the Harz Mountains: flooding. Waterwheels and horse-powered treadmills soon proved in-adequate for such a monumental task. It was left to two devices that harnessed some of the knowledge that had come from the explosion of science in the second half of the seventeenth century, led in London by the Royal Society and in Paris by the Academy of Sciences. The inven-tor of the first such device was the British military engineer Thomas Savery, He developed a simple steam pump that could harness some of the energy of burning coal and turn it into mechanical energy. In his device, steam entered a cylinder and was condensed, producing a vacuum that pulled water up into it. More high-pressure steam then entered the cylinder to expel the water and pump it upward. Savery's device was cheap to construct but needed high-pressure steam and had to be built deep down in the mine close to the water level, so it proved impractical.

The second device, the atmospheric engine, was first proposed by the Frenchman Denis Papin, who had oscillated between London and Paris, working alongside such luminaries as Robert Boyle and Robert Hooke on the age's favorite experimental toy, the air pump. Papin's ideas were turned into a final practical form by a Devonian ironmonger Thomas Newcomen. His atmospheric engine used the same principle as Savery's device, but used the vacuum in a cylinder not to raise water but to pull a piston down and to harness its movement to drive a water pump. This device was much larger and more complex and needed to be built as part of a huge framework, but it could be sited at the surface of a mine and was capable of further development. Newcomen's atmospheric engines, which commonly managed to achieve outputs of around ten horsepower (7.5 kilowatts), consequently became more and more common in coal mines throughout the country. Their main drawback was their inefficiency, since only 0.5 percent of the heat energy from the coal was converted to mechanical power. Fortunately, since most engines were used in coal mines, this was not a pressing problem since they could be fed with waste coal. Around six hundred had been built by 1775, and they even started to be used to drain the tin mines of Cornwall.

By the middle of the eighteenth century, therefore, Britain was poised to become fully industrialized. It fed its growing population with a productive agricultural system. It powered its manufacturing with water mills that could be used all year round and heated its houses and industrial furnaces with coal. It linked its industries to their raw materials and markets via a growing network of canals. On the high seas it protected its merchant fleet with the world's most powerful navy, which also kept open the markets that it was cultivating around the world, most notably across the Atlantic in its Caribbean empire and its colonies in North America.

Sad to say, though, almost none of this development was documented by British artists as the Dutch painters had showcased their country's Golden Age. British eighteenth-century culture was dominated by the aristocracy and landowning classes, who were obsessed with Italian culture, so the architects, painters, and opera composers copied Italian models. They seemed to want to maintain the illusion that Britain was still an arcadian paradise. Not until Joseph Wright of Derby in the late eighteenth century did a painter emerge who was willing to highlight British science and industry, notably in his depictions of scientific demonstrations of the air pump and orrery, and his scenes set in blacksmiths' forges. To heighten the drama of his scenes he lit his works with brilliant chiaroscuro effects, reminiscent of the works of Caravaggio. And it was Wright who painted the portraits of the generation of Midlands industrialists who would finally transform Britain into the workshop of the world and give birth to that mainstay of industrialization, the factory.

Chapter 14

POWER FOR PRECISION

In a year of famous declarations, 1776, perhaps the most important for the future of humankind was not the one published by the founding fathers of the USA, but the one spoken by the British industrialist Matthew Boulton. On March 4 he showed the Scottish diarist and biographer James Boswell around his Soho works near Birmingham. The climax of the tour was the engineering workshop, where Boulton and Watt's pioneering steam engines were being assembled. You can imagine Boulton standing proudly erect with his thumbs thrust into his waistcoat pockets as he proclaimed, "I sell here, sir, what all the world desires to have—*power.*"

It was a startling boast to make in a country where a man's economic and political power was inextricably linked to his possession of land, inherited wealth, and most of all to his aristocratic birth. But Boulton was speaking at a time when Britain's wealth and power were starting to be based not just on its agriculture or its growing empire in the Caribbean, North America, and India, but on its manufacturing capability. And the story of the crucial role that mechanical power played in the expansion of Britain's manufacturing goes to the very heart of the journey that Eurasian civilization took toward creating our modern industrialized consumer society.

Early-eighteenth-century Britain was certainly able to produce more food and mine more coal than ever before, so that it could feed an expanding workforce and expand its heavy industry. However, on their own these could not help Britain raise the productivity of many of the manufacturing trades on which it had traditionally relied, particularly the dominant textile industry. In such complex processes as producing cloth, what limited the productivity and output was not the power available, but the time it took workers to undertake the many skilled tasks.

Writing in the same year as Boulton's famous speech, 1776, this was already clear to the pioneer of economics Adam Smith. In perhaps the most famous passage of his magnum opus, *The Wealth of Nations* (and probably also the most read, since it is found on page one of his five-hundred-plus-page book), Smith extolled the virtues of the division of labor. If you split up a task such as the manufacture of pins into many component parts, you can speed up production manifold; the division of labor enables individual workers to specialize and excel in just one task and eliminates the time they would waste moving between tasks. By the time that Smith was writing, division of labor of this sort had started to become a notable feature of British industry. Josiah Wedgwood was using it, for instance, to speed up the production of ceramics in his Etruria factory in Staffordshire; individual workers there specialized in the separate roles of throwing, turning, handling, and decorating his pottery, while others mixed the slip. Meanwhile his fellow member of the Lunar Society Matthew Boulton was using the same technique in his Soho works, which made metal "toys" such as buttons and buckles. But division of labor alone could not have produced the massive rise in industrial output and productivity that occurred in Britain's textile industry during the second half of the eighteenth century. Fortunately, you don't have to read much further in Smith's book, only to page two, to discover the major factor that further increased the output of finished cloth: mechanization.

The manufacture of textiles was particularly ripe for mechanization because of the sheer complexity of the tasks that are needed to convert microscopic fibers of wool, linen, cotton, or silk into finished garments. Textile workers had long been specialists in single parts of the industry—carding, spinning, weaving, and fulling—so there was some division of labor. But all of these processes demanded high levels of skill and dexterity, and they were all time-consuming. And long before the middle of the eighteenth century, most of the processes had already been speeded up by the invention of human-powered machines.

By medieval times, European women had already transformed spinning by using that most sophisticated of preindustrial machines, the spinning wheel. First imported from India, a spinning wheel works by mounting the spindle horizontally on an axle and spinning it around using a belt coming from a much-larger flywheel. The spinner teases out the fibers between their fingers and lets it be pulled through a ring on another structure rotating about the spindle, the flyer, which provides the twist before the thread winds onto the spindle. The beauty of the European spinning wheel was that the process was continuous and used little energy, since the spinner could keep the wheel rotating using a treadle, which left both their hands free to tease out the fibers. This doubled the speed at which women could produce thread compared with the old drop spindle.

Weaving—building up fabric by crisscrossing one set of threads at right angles across another set—had also been facilitated for thousands of years by the use of looms. At first these consisted merely of frames to hang the threads on, but by early modern times looms had been developed into what were effectively sophisticated foot-powered machines. The weaver pushed two treadles up and down, raising and lowering two sets of alternating warp threads. Between these actions, the weaver passed a bobbin of weft thread back and forth from one side of the loom to the other and firmed the new weft thread into the cloth by pressing it with a wooden board, the batten. The process worked well, but the length of weavers' arms limited the width of cloth they could make and spoiled their posture; since they had to lean forward to pass the bobbin across the cloth, weavers were notoriously round-shouldered.

In 1589, the English inventor William Lee had even devised a "knitting frame," which could speed up the complex process of knitting, in which threads are looped and knotted around themselves using convoluted movements of a set of knitting needles. By the eighteenth century, a highly successful knitting industry had been set up by Huguenot settlers in the Midlands towns of Leicester and Nottingham.

For all the ingenuity of the devices that the textile workers were using in the middle of the eighteenth century, however, textile work was still effectively a set of cottage industries. Though the workers were often employed on piecework and rented machines from wealthier merchants, they carried out the work independently in their own homes, and the work was highly skilled and powered by hand. By the start of the eighteenth century the British textile industry employed hundreds of thousands of people, especially in the Midlands and the north of England. The progressive removal of human craft skills from these industries, the application of water- and steam-powered machinery, and the concentration of the machinery in large factories or mills marked a dizzying rise in British productivity. Within a century, the industry had been transformed beyond recognition. The ingenuity of textile workers, especially those from the cotton-manufacturing county of Lancashire, drove mechanization, but the reason change occurred in the eighteenth century and not before was because of the development of a new legal framework to reward innovation: patents.

Traditionally, rulers of states would gain little by helping their citizens improve their living standards. After all, a population that was better fed and equipped might have more leisure to consider alternative forms of government and more opportunities to rebel against their present rulers. It was better to involve them in wars that would distract them and might increase the prestige of the ruler. Most laws were therefore set up to maintain the status quo, and one major building block of a state were its guilds: self-governing trade organizations that limited the numbers and activities of craftsmen. With their lengthy apprenticeships, the guilds kept up the standard of work but acted as a brake on innovation; they ensured that traditional techniques were passed on unchanged from generation to generation. It is not surprising, therefore, that technological progress was so slow and that living standards in Europe barely changed over the centuries, from ancient times right up to the sixteenth century.

Patents helped end this stranglehold. They are agreements that allow inventors exclusive rights to their invention for a number of years. Armed with a patent, inventors were given time to achieve both profit and prestige. Following their first introduction in the Venetian Republic in 1474 the idea spread, reaching England in the early seventeenth century. Under pressure from the people and parliament, James I introduced the Statute of Monopolies in 1624. This stated that the king could only grant monopolies for a fixed number of years and only then to new inventions. The law was further strengthened at the start of the eighteenth century so that patents had to give complete specifications for their inventions. The starter gun for the innovations that were to transform the textile industry was fired in weaving. In 1733 the Lancashire weaver John Kay patented an innovation that could double the speed of cloth production and enable weavers to make much wider pieces of cloth: the flying shuttle. This was essentially just a streamlined box on wheels that held a bobbin of weft thread, which was mounted in a groove of the batten board of the loom. To move the weft back and forth all the weaver now needed to do was to flick the shuttle back and forth into and out of boxes on either side of the loom, using an arrangement of strings. Weaving became much faster, if noisier, and weavers' postures improved.

This change stimulated advances in other areas of textile production. Using conventional spinning wheels, spinners could no longer meet the demand for thread from the faster-working weavers, and in the 1760s this led inventors to devise machines that could produce more thread using less labor. The main difficulty was how to replace the skill and dexterity of spinners' fingers: the way they drew and evened out the fibers from the ball of raw cotton, before their spinning wheels twisted it and wound it onto the bobbin. The first and best known of the machines, Hargreaves's spinning jenny, was basically a multiple spinning wheel that spun eight or more threads at a time. To mimic the actions of a spinner, the machine simply pulled the cotton rovings between two bars before twisting them and winding them onto rotating bobbins. All the spinner had to do was to turn the wheel to keep the machine moving and mend any threads that broke. This certainly speeded up thread production, but the smoothing arrangement was crude, so the thread produced was weak and inconsistent; it could only be used as the weft in a loom. So though two hundred thousand spindles were making thread using spinning jennies by 1788, another machine was needed to speed up the production of warp thread.

The answer was the water frame, patented by a Preston wigmaker, Richard Arkwright. This machine mimicked the way a spinner manipulates the fibers much better, by running the roving through three sets of rollers, each of which spun faster than the previous one, so the machine gently stretched the material before twisting it. Arkwright's machine produced much stronger, more even thread than the spinning jenny and was followed in 1785 by another machine that combined the action of the other two designs: Samuel Crompton's mule. This machine operated in two stages, rather like a spinner using a drop spindle. In the first stage, rollers drew out the roving before the material was twisted, and the thread was pulled out by a wheeled carriage that rolled out away from the main structure. In the second stage the carriage returned, and the machine wound the spun thread onto the spindle. The result is an impressive and mesmerizing spectacle. Mules, often spinning hundreds of threads at a time, were to dominate the spinning industry well into the twentieth century.

Spinning machinery was soon joined by machines that took over other aspects of textile production. For instance Arkwright also devised a carding machine, which converted raw cotton into rovings for his spinning machine. This machine also used rollers, but in this case gigantic slow-moving ones, covered in mats of hooked teeth that teased out the fibers and removed the dirt as the rollers sheared past one another, to produce a fine stream of clean and perfectly aligned cotton fibers. The American industrialist Eli Whitney used a similar principle to create yet another machine, the cotton gin, which removed the seeds from raw, recently harvested cotton. This greatly sped up production on the American cotton plantations, improving their competitiveness and enabling them to expand.

As the spinning technology developed, hand weaving once again became the limiting process in textile manufacture. This eventually led inventers to develop looms powered and operated by machines, not people. They devised levers and cams to move the warp threads up and down, and to push the shuttles back and forth across them. They even came up with ways of automating the manufacture of cloth with complex patterns such as brocade and damask. The French inventor Joseph-Marie Charles Jacquard invented the Jacquard loom, which used cards, rather like those that inputted information to early mainframe computers, to determine which of many shuttles was moved across the fabric. Finally, once the cloth had been woven, other machines were devised to stretch it, roll it, and tease out fibers to raise the nap on velvet and moleskin.

Knitting machines were also improved to increase the range of materials and garments they could produce. For instance in 1758 the Derby inventor Jedediah Strutt developed an attachment to knitting frames, the Derby Rib machine, which enabled him to produce rib stitch as well as plain knitting. This meant that he could make the whole of a knitted garment, not only the main body, but also the clingy cuffs around the wrists that keep your arms snug and the gripping tops of socks that hold them up. And in 1816, the French immigrant engineer Marc Isambard Brunel produced a circular knitting machine that could make seamless socks and stockings that did not need to be sewn together.

As the textile machines grew larger, more complex, and more costly than simple spinning wheels or looms, only wealthy capitalists or entrepreneurs backed by banks could afford to buy them. The machines were set up in large factories or mills, where the formerly independent craftspeople worked as dependent employees. But though the rollers, cams, levers, and wheels moved more precisely than a worker's fingers, they were much heavier and encountered far more friction than our bodies. Consequently, although the machines worked faster, they required more power to operate. A single worker could just about generate enough energy to operate a single spinning jenny with ten spindles, for example, but Arkwright's water frames, with their greater number of moving parts, would have been beyond them. Indeed, when Arkwright set up his first spinning mill in Nottingham, he found that even horses harnessed to a treadmill could not meet his needs. He raised more funds and in 1771 set up a water-powered mill, possibly the world's first factory, in the village of Cromford on the River Derwent in Derbyshire. This enterprise became the blueprint for the development of all subsequent textile mills.

The result of all this mechanization was a dramatic rise in output and productivity, creating an economic miracle that turned the English textile industry into a wealth-creating force that transformed the north of England and the East Midlands into industrial powerhouses. It drew in raw materials such as cotton, flax, and silk from distant countries such as India and the United States, transformed it into cloth and garments, and exported these all around the world, even to the countries from which it had obtained the raw materials. But mechanization had even more profound effects than just raising the productivity of one industry; it also transformed the way people work, the structure of society, and even our understanding of power and energy.

To get an idea of the impact of the textile industry, it is perhaps best to visit one of the superb working museums set up in former textile mills that showcase their work. The overall impression of the machinery is one of wonder at its speed and precision, and I remember being almost hypnotized by the back-and-forth motion of the looms when, aged nine, I visited my first textile museum at the Trefriw Woollen Mills in Betws-y-Coed, North Wales. I have been visiting mills ever since, especially since I moved to the marvelous north of England. But perhaps the mill that best displays not only the machinery, but also the geographical, social, and intellectual effects of mechanization, is Quarry Bank Mill, near Manchester. A vast multistory brick edifice planted incongruously in the idyllic Bollin Valley, its cavernous interior hosts long banks of spinning and weaving machines that clatter and hum as they move inexorably back and forth, driven by a huge pounding waterwheel and a hissing, roaring steam engine. All around the valley there are other ancillary buildings: lines of tiny terraced cottages that housed the labor force for the mill on the plain above, and an "apprentice house" or orphanage nearby to provide child labor.

The most immediately obvious aspect of this industrial campus is its isolation. Rather than being sited in a big city or town, mills had to be set up beside a convenient source of power: a river or stream to drive a waterwheel. Mills had to be new builds, and proprietors also had to build living quarters for their workers to create an industrial village. The next impression you receive is the sheer size of the mill, larger than a cathedral and a far cry from the tiny isolated cottages where women had worked at their spinning wheels by the fire while their husbands wove cloth on their looms in the well-lit attic above. The size and ordered arrangement of the machines and of the cottages that surround the mill also point to the profound social consequences of mechanization. The mill owners provided the capital to build the mill and purchase the machines. They guaranteed the workers regular work and decent wages, but in return the workers had to commit to rigid working hours and strict rules. And the work itself was de-skilled; instead of spinning thread themselves, the workers merely tended the machines. Rather than being independent craftspeople, workers were now hired hands.

But the most dominant aspect of mills was their huge demand for mechanical power. Not only were the machines profligate with energy because of all the rollers, cams, levers, and wheels that had to be moved around, but they had to be linked to the waterwheel via a series of gears, cogs, shafts, and belts, which wasted a lot of the power even before it could reach the machines. Consequently, though producing cloth was more efficient in labor, it was much less efficient in power consumption. It took far more power to produce a garment using machinery, and the output of a mill was limited not by the input of human skill but of mechanical power. At Quarry Bank the massive power requirement can be seen simply by looking at the size of the waterwheel. With a diameter of thirty-two feet (9.7 meters) and width of twenty-two feet (6.7 meters), it could produce a massive 120 horsepower (90 kilowatts), many times more than the wheel of a typical grain mill and the equivalent of fifteen hundred men!

With the bottom line at stake, industrialists suddenly realized the crucial role that mechanical power played in their enterprises, which explains Matthew Boulton's boast. Everyone suddenly wanted power. It became important to work out ways to supply more mechanical power, to measure power more precisely, and to define what power actually is.

The most immediate need for the mill owners of the time was to choose the best design for their waterwheels; they needed to know what sort of wheel to install, how big they needed to make it, and how much power it could produce. At the time engineers did not even know which was the better type of waterwheel: undershot or overshot. This led a brilliant young engineer from Leeds, John Smeaton, to carry out a series of experiments on model waterwheels, which aimed to determine how efficient the different designs were. In 1759, Smeaton showed that overshot wheels were twice as efficient as undershot ones, around 60 percent, compared with 25–30 percent, and he produced guidelines to inform a mill owner how much power any given waterwheel could produce, and the optimal design for any site.

Meanwhile, mill owners also needed an additional power source that could help them maintain production during droughts in summer, or extreme frost in winter, weather conditions that could put their waterwheels out of action. An alternative already existed, as we saw in the last chapter, in the form of Newcomen's atmospheric engines, but they were inefficient and their high coal use made their use uneconomic away from coal mines. Fortunately, though, by the middle of the eighteenth century, engineers had started to improve their performance. Smeaton, for instance, reduced energy losses by insulating the cylinders of his engines and managed to double their efficiency to around 1 percent.

The game-changing breakthrough, however, was made by the Scottish engineer James Watt, who added a separate condenser to his atmospheric engines, dramatically reducing heat loss from the cylinder, and further doubling their efficiency to 2 percent. Watt also devised a way to enable his engines not to pump water, but to drive machinery, by reversing the beam arrangement that engineers had previously used to convert rotational to reciprocating motion. The piston rod of his engine was linked by Watt's ingenious "parallel action" to one end of a rocking beam, while the other end drove a crank that was linked to a flywheel via a set of gears. For the first time a steam engine could power rotating machinery, not just a pump.

These developments in powering textile machinery had another important impact, this time on science. As engineers built more and more powerful waterwheels and steam engines, they needed a way to measure the power they could produce. Not only would this enable a mill owner to calculate how many machines he could run from them, but it would be a marketing tool; it would emphasize just how good the engines were compared with alternative power sources. To do this they came up with the concept of work, the ability of a machine to lift a given weight of water to a given height. Engineers could measure what they called the power of their engines—the rate at which they could perform work—by running their engines against a brake. The power an engine could produce was the force applied to the brake multiplied by the speed at which the wheel rubbed past it. These practical engineers had come up with a key idea that was to lead to the modern concept of energy. They were fifty years ahead of the scientists of the day, who were stuck thinking only about forces in the old Newtonian way, in relation to acceleration and momentum. The only problem was the esoteric nature of the unit they used, foot-pounds per minute.

To market the power of their machines in a way that was intelligible to mill owners and to the general public, engineers had to come up with a more intuitive unit of power, and they achieved this by comparing the performance of machines with that of the horses that they were replacing. Smeaton sensibly undertook trials of horses operating a scoop wheel and came up with a figure for power output of twenty-two thousand foot-pounds per minute per horse. However, the figure we use today was that put forward in 1782 by Watt. Having discussed the matter with millwrights, Watt learned that over a working day a horse could walk two and a half times per minute around a twenty-four-foot-diameter circular treadmill, while pulling with force of 180 pounds. This gave a round figure for the average power a horse could supply as thirty-three thousand foot-pounds per minute. Despite this estimate's being rather high, especially for the small horses of the day, Watt's is the figure we still use for horsepower. One horsepower equates to 735 watts in the modern universally agreed scientific SI (Système Internationale) units.

For the first time in history, the concept of mechanical power had been defined, its importance had been acknowledged, and it became a matter of common speech. But the technological advances made by the eighteenth-century engineers only marked the start of the expanding role that machinery and mechanical power was to play in the history of the world. Though the waterwheels were huge and powerful, by the end of the eighteenth century, there were still few water-powered textile mills in Britain, 210 in 1788 and around 1,000 in 1800, giving a total capacity of around one hundred thousand horsepower (73 megawatts). This was only a fraction of the power available from the working population of 8.9 million people, around six hundred thousand horsepower (440 megawatts). And despite the importance historians place on the significance of Watt's steam engines, by 1800, Boulton and Watt had only sold around four hundred of their rotative engines, and only around fifteen hundred steam engines had been installed in total in Britain. Since each engine typically produced only around ten to fifteen horsepower, this gave the tiny capacity of only fifteen to twenty thousand horsepower (11–15 megawatts), around a fifth of the power from water and just 2–3 percent of that available from human labor.

And with the mills scattered around the countryside, England remained a largely rural country, so the imprint of industrialization is absent even in the works of its champion Joseph Wright of Derby. Most of his landscapes show the picturesque scenery of the Peak District, familiar to viewers of historical dramas such as *Pride and Prejudice*. His portrait of the industrialist Richard Arkwright does show him in front of his mill, but since this was built in the popular Palladian style, it resembles a country house more than a factory. Even the mill workers retained some of their country ways. The workers' cottages at Quarry Bank have large gardens where the employees could grow much of their food, including for the first time that most productive and nourishing New World crop plant, the potato. The expansion of power that would fully industrialize and urbanize Britain still awaited three technologies that were just beginning to emerge at the end of the century.

Chapter 15

POWERING AN INDUSTRIAL NATION

One event more than any other highlighted the progress of industrialization in the first half of the nineteenth century and showcased Britain's dominant position and technological superiority: the Great Exhibition of 1851. Britain was probably the only nation that could have staged the event, and its vast scale and audacity staggered the world. It was a far more effective display of political and economic power than the most extravagant triumphs of ancient Rome, since it not only presented what the British had plundered from their colonies, but what it had produced itself. It was the ultimate example of what has come to be known as soft power.

Since its scale still seems Brobdingnagian today, it is worth presenting some of the mind-boggling statistics. The building that hosted what was essentially the world's first trade fair—the Crystal Palace—was an engineering marvel of iron, glass, and wood that covered almost twenty-five acres (10 hectares) of London's Hyde Park. Being over six hundred yards (550 meters) long, almost 160 yards (146 meters) wide, and with a maximum height of over 130 feet (40 meters), it was three times the size of St Paul's Cathedral, yet it was built in under five months, and after the end of the exhibition it was taken down and set up at a permanent site elsewhere in the capital, like a huge tent. During the summer of 1851, the building housed over one hundred thousand exhibits from all around the world, the majority of the exhibitors, seven thousand, being from Britain itself or its colonies, whereas only six thousand were from the rest of the world. The exhibition was visited by six million people, an astonishing number from a country with a population of only 21 million. And despite the fact that the entrance fee fell to just a shilling by its end, the exhibition made a substantial profit. The organizers subsequently bought land just south of the park and built a massive new cultural campus, comprising the Natural History, Science, and Victoria and Albert Museums, the Imperial College of Science and Technology, the Royal Academy of Arts, Music and Organists, and the Royal Albert Hall.

None of this would have been possible with the technology available fifty years before. The Crystal Palace could not have been built; the exhibits could not have been transported to the site; and the visitors could not have traveled there in sufficient numbers. The exhibition owed its existence to the ability of British engineers to produce and harness undreamed of amounts of mechanical power. They transformed the nation in the first half of the nineteenth century, giving the country unprecedented economic and political power.

Despite the huge attention paid to James Watt—he is widely feted as the father of the steam engine—his machines were underwhelming devices, even by late-eighteenth-century standards. They could never have powered the world of the mid-nineteenth century. As we saw in the last chapter, they produced a mere fifteen to twenty horsepower (11–15 kilowatts), so they were far less powerful than large waterwheels, and they had a miserably low efficiency of just 2 percent. They were linked to machinery using inefficient sun and planet gears, and they were huge and heavy. Moreover, they could never be used in a moving vehicle since they needed a constant cold water supply for their condensers.

It was not all Watt's fault. At the time, boilers had to be made from cast iron, which could not withstand large tensile forces; they would have burst if Watt had raised the temperature and pressure of the steam inside much above atmospheric pressure so it could push the piston as well as pull. Watt's engines had to have huge cylinders to harness the power of atmospheric pressure as the steam was condensed within them, producing a vacuum. Their cast-iron beams also had to be thick to withstand the bending forces to which they were subjected; most weighed upward of six tons. The enormous structures gave his engines a power-to-weight ratio of less than one horsepower per ton, much less than a horse!

The rise to greatness of Britain's industry in the first half of the nineteenth century was instead based on three new technologies that have gone largely unheralded by popular historians: wrought iron, high-pressure steam engines, and the hydraulic press. Working symbiotically, improvements in each technology helped spur on developments in the others. They enabled engineers to generate and harness the vast amounts of power and the gigantic forces that they needed to transform Britain into a country that was industrialized, well connected, and overwhelmingly urban.

The first of the three technologies to be developed was the new wonder material of the early nineteenth century—wrought iron. This material enabled the construction of the Crystal Palace; the building of the engines and machines that powered the industry it showcased; the building of the locomotives and the rail infrastructure that transported people to and from the site; and the building of the steamships that brought the exhibits from around the globe. Wrought iron ushered in the heroic age of engineering and of the celebrity engineers who would construct a new nation.

We saw in chapter 13 that smelting iron ore with coke instead of charcoal had allowed ironmasters to dramatically raise the output of cheap iron. However, though iron could be molded into a wide range of complex shapes, because of its high carbon content this "cast iron" was extremely brittle, so it could not be used to make structural components such as beams, connecting rods, or chains. To toughen the iron, engineers had to melt the cast iron down again and hammer it into shape to produce rod iron, as blacksmiths had done for centuries, but this took time and negated the cost benefits of making cast iron in the first place. By 1740 a Sheffield ironmaster, Benjamin Huntsman, had developed a method of making an even better material—steel—by heating iron with coke to a temperature of 2,900°F (1,600°C), melting it, and allowing a flux of sand or glass to react with and remove the carbon. The superior quality of the "crucible steel" he produced enabled the Sheffield steel industry to expand, rapidly increasing its output from two hundred tons per year in 1700 to ninety thousand tons per year a century later. Sheffield became a city of "little mesters," whose small workshops crafted all manner of tools and cutting implements, from tableware to scissors, knives, and the blades of scythes. But crucible steel was only made in small amounts—each crucible held around thirty-three pounds of steel—so it was far too expensive to use in larger pieces of machinery, let alone in vast engineering works.

The answer to the problem—for a hundred years at least—was wrought iron. In 1784, the Welsh ironmaster Henry Cort developed a technique that could quickly and cheaply produce large quantities of this material, which was as strong and tough as traditional rod iron but much cheaper. In Cort's patented "puddling" technique a lump of cast iron was melted in a furnace so that it formed a puddle on the floor. The puddler then stirred it with a long rod, causing the carbon in the metal to react with iron oxide strewn over the floor and to bubble off as carbon monoxide. This process raised the melting point of the iron, so that it started to solidify as the puddler continued to roll and fold it, incorporating slag fibers into the mix. Finally the puddler lifted the lump of metal out of the furnace and shaped it into plates or rods by squeezing it repeatedly through a series of rollers. A single puddler could produce 220 pounds (100 kilograms) of wrought iron a day, making it far cheaper than crucible steel or rod iron, while the slag fibers protected it from corrosion.

The new material was soon being exploited by engineers working for the Cornish tin-mining industry, enabling them to produce the second technology: the high-pressure steam engine. At the beginning of the nineteenth century, the miners were using Watt's improved steam engines to pump water out of their mines, but these machines used a lot of coal, which was much more expensive in Cornwall, far from any coal mine, than it was where the textile industry was based, in Lancashire and Yorkshire. To produce more powerful and more efficient steam engines, Cornish engineers such as Richard Trevithick riveted wrought-iron plates together to produce cylindrical boilers that could withstand pressures of several atmospheres. These provided high-pressure steam for a new generation of Cornish beam engines that could provide more power, while being smaller and using less coal.

To make use of all the available energy in the steam and to raise the efficiency further, they developed two new techniques. One was to close off the valve letting steam into the cylinder when the piston had only moved part of the way along it, so that the expansion of the high-pressure steam powered the remainder of the stroke. The alternative to "expansion working" was to construct compound engines that ran the steam through two or three sets of cylinders of increasing size, making use of steam at high, medium, and low pressures. The power and efficiency of engines progressively increased until by the 1840s Cornish engineering firms were producing huge pumping engines that ran at pressures of over four atmospheres, produced hundreds of horsepower, and reached efficiencies of over 10 percent, five times that of Watt's engines. The largest of all was the Cruquius pumping engine, built in 1849 on the outskirts of the Dutch city of Haarlem. This monstrous machine with its 144-inch (3.6-meter) diameter pistons developed around three hundred horsepower (220 kilowatts). This one machine drained seventy square miles (180 square kilometers) of lake and swamp in three years, creating prime new agricultural land around twenty times faster than all the windmills of the Dutch Golden Age combined had managed in the seventeenth century.

Engineers soon found that the cylinders of their engines did not need to be oriented vertically. Once precision machine shops had developed metalworking planes that could produce smooth runs for the piston rods, engineers started producing engines with horizontal cylinders. These machines did not stress the floor of the building so much, and they could be directly coupled to a flywheel by a simple crank, so there was no need for a heavy beam. This meant that the engines could easily be run at high speeds. Small, cheap horizontal engines replaced massive beam engines in textile mills, and by the 1850s larger models could produce over five hundred horsepower (370 kilowatts). Weighing around one hundred tons, they had a power-to-weight ratio around five times higher than Watt's engines. Industrialists could power their mills cheaply and effectively using just a single high-pressure steam engine. And because of the efficiency of their machines they could site their mills wherever was most convenient; they did not need to be built beside fast-flowing streams or close to coal fields.

High-pressure steam engines also enabled engineers for the first time to construct machinery out of iron, which is a far more rigid and stable material than wood. In the first half of the nineteenth century, a new breed of mechanical engineers developed a series of lathes, sanders, planes, milling machines, and drills that could work iron in the same way that traditional machines had worked timber. Engineering giants such as Joseph Bramah, Henry Maudslay, James Nasmyth, and Joseph Whitworth set up steam-powered machine shops that produced metal components with higher and higher levels of precision. They made the machines that were to drive industrial growth and the machine tools that would make the next generation of machines. Within one of these machine shops, that of the locksmith and inventor of the flush toilet, Joseph Bramah, the third innovation was hatched: the hydraulic press.

From ancient times, people had been using a variety of mechanical devices to magnify the forces they could produce with their own muscles. Farmers had been using levers or screw presses for hundreds of years to crush olives and grapes. Ships had been fitted with capstans to help sailors haul in their ropes. And military engineers had fitted their ballistas and crossbows with winches to pull back on their strings. The problem with all these devices was friction. Bramah's flash of genius was to use Pascal's principle—that the pressure in a body of fluid is constant throughout—to produce a near-frictionless hydraulic lever. Bramah's hydraulic press had two pistons, a narrow one and a broad one, connected by a tube of water. Pressing down with a small force on the narrow piston generated a much larger force on the broader one. At first engineers used hydraulic presses to magnify muscle power, to bale cotton or extract oil from linseeds. However, when the presses were linked to steam-powered pumps, they could generate much larger forces; machinists could even use them to shape metal components: bending them, cutting them, and perforating them. I remember being mightily impressed when I first saw a hydraulic press in action, distorting a flat sheet of aluminum into a bowl shape to produce one-half of a beer barrel, just as you or I would press pastry into the depression in a baking tin. With a hydraulic press metals become putty in our hands.

An early hand-powered hydraulic press being used to bale cotton.
Later versions were much larger and powered by steam engines.

In the first half of the nineteenth century, the holy trinity of wrought iron, high-pressure steam, and the hydraulic press transformed manufacturing, but to the layman its most obvious effect was on transport. Perhaps the greatest benefit from high-pressure steam was the development of steam engines that could power their own movements—locomotives. Not only were high-pressure engines smaller and lighter than traditional beam engines, but they no longer needed a separate, stationary condenser with its own water supply; the steam would escape from the cylinder at the end of the stroke because of its high pressure. Engineers could even put the escaping steam to good use, blowing it across the entrance to the engine's chimney to help draw hot gases through the boiler, and producing the chuffing sound beloved of steam enthusiasts.

The pioneer of steam locomotion was the Cornishman Richard Trevithick, who produced a whole series of steam locomotives, including the *Puffing Devil,* the first steam-powered road vehicle, and *Catch Me Who Can,* the first locomotive to run on rails. However, early steam locomotives were too heavy for roads, and their cast-iron rails all too often broke under their weight. Consequently, railways only started to expand when engineers had developed a method to produce rails out of wrought iron. They used steam engines and hydraulic presses that moved lengths of hot iron back and forth between sets of rollers to create rails with the necessary I-shaped cross section. Steam rolling mills could also produce the plate to make the engine boilers themselves and all manner of engineering components. To get an idea of the sheer power of these devices, the best place to go is perhaps the Kelham Island industrial museum in Sheffield, where you can still see one in action. The River Don Engine generates some twelve thousand horsepower (8.8 megawatts) and changes direction at will. It was developed in the late nineteenth century to roll steel plate, such as sixteen-inch (40-centimeter) armor for battleships and later the blast floors of nuclear power stations.

Wrought-iron rails could support ever larger, more powerful loco-motives, and with the advent of the Stockton-to-Darlington railway of 1825, and the Liverpool-to-Manchester line in 1830, the world's first passenger railways, steam-driven travel expanded rapidly. The length of railway line in Britain increased from just ninety-eight miles in 1830 to fifteen hundred miles in 1840 and sixty-six hundred miles in 1850. By 1850 there were also seventy-three hundred miles of railway on conti-nental Europe, while in the United States railroad growth exploded even faster so that there were twenty-eight hundred miles of track in 1840 and twenty-nine thousand miles in 1850.

The holy trinity also helped build another essential component of the rail infrastructure: bridges. Because of its brittleness, cast iron was never suitable to build long railway bridges. So after several disasters, notably the collapse of Robert Stephenson's cast-iron truss over the River Dee in 1847, engineers turned once again to wrought iron, which en-abled them to build ever-longer, more spectacular structures. From the 1820s onward, engineers such as Samuel Brown and Thomas Telford had exploited the high tensile strength of wrought-iron chains to build spectacular suspension bridges for road traffic, most notably Brown's 440-foot (135-meter) long Union Chain Bridge over the Tweed, and Tel-ford's 577-foot (176-meter) span bridge over the Menai Strait between Anglesey and the Welsh mainland. However, suspension bridges were too flexible to cope with the heavy moving load of a railway locomotive. So when Robert Stephenson was designing the first long rail bridges, he made them out of wrought-iron beams. His 1847 Britannia Bridge across the Menai Strait was composed of two huge wrought-iron tubes, each some four hundred feet (125 meters) long, through which the trains could run. Though successful, this solution used a lot of iron and was expensive, so in his 1849 High Level Bridge over the River Tyne at Newcastle, he chose to bridge the 125-foot (38-meter) spans between each support with wrought-iron bowstring trusses. Most subsequent rail bridges were con-structed using trusses composed of crisscrossing wrought-iron girders.

But the structural design and the material Stephenson used were not the only novel aspects of his bridges. He also used novel techniques to build them. In the early 1840s the Scottish engineer James Nasmyth and the French engineer François Bourdon developed the steam hammer. This was essentially a high-pressure steam engine that raised and lowered a heavy piston, which acted like a huge mallet. Metalworkers quickly put steam hammers to use in their forges to shape large castings such as the driveshafts of steam engines, and they were developed into machines to rivet wrought-iron plates together. Nasmyth also adapted the machine for work in the field, producing the world's first steam pile driver, which could produce a blow thousands of times more powerful than a man with a mallet. Stephenson used the machine to install the piled foundations of the High Level Bridge and of another bridge on the London-to-Scotland line, the Border Bridge, a long masonry viaduct across the River Tweed at Berwick. Engineers could also combine steam engines and hydraulic presses in the field. Stephenson used steam-powered hydraulic rams to raise the fifteen-thousand-ton tubes of the Britannia Bridge up to their piers.

A steam hammer in full work (1883) from a painting by James Nasmyth.

By the time of the Great Exhibition in 1851, Stephenson's bridges had completed the east-coast mainline between the Scottish capital, Edinburgh, and London, reducing the travel time between them to just over twelve hours. His Britannia Bridge also completed the rail link between the ferry terminal at Holyhead and London, thereby speeding up links to the Irish capital, Dublin. By 1851 it was possible to travel between these two capitals in just seventeen hours. Most people from England could reach London even more quickly. By 1851 steam locomotives had reached speeds of up to 78 mph (125 kph), and most trains traveled at an average speed of around 40 mph (64 kph), so travel times to London from Newcastle were just over eight hours, from Manchester just over five, and from Birmingham, just three hours. It was the speed and low cost of train travel that enabled so many people to take economy excursions to London to see the Great Exhibition.

If steam locomotives speeded up transport on land, the effects of steam power and wrought iron on ocean travel were, if anything, even more transformative. Before the nineteenth century all sea travel was by wooden sailing ships, which were small, slow, at the mercy of the winds, and leaked chronically between their timbers. Between 1800 and 1850 ship design underwent a total revolution. At first, iron was used solely as armor for wooden warships, but marine architects gradually learned how to borrow the techniques used by boilermakers. They built a new breed of iron ships by riveting wrought-iron plates together into huge waterproof hulls, reinforced inside by bulkheads and stringers. And they augmented and finally replaced the motive power of sails with larger and larger steam engines, which drove huge paddle wheels and propellers. By 1850, the British engineer Isambard Kingdom Brunel had launched the first screw-powered iron ship, the SS *Great Britain*, the largest passenger vessel in the world at thirty-seven hundred tons. Powered by two five-hundred-horsepower (370-kilowatt) steam engines weighing 340 tons in total, it could transport seven hundred passengers over the Atlantic in just twelve to fourteen days. Steamships made it possible for industrialists from around the world to transport their exhibits to London.

Finally, the holy trinity transformed architecture and made the building of the Crystal Palace itself possible. From the beginning of the nineteenth century, mill owners had been using cast- and wrought-iron frames to hold up the floors of their buildings and support the heavy iron machinery within them. The first, William Strutt's North Mill in Belper, Derbyshire, was completed in 1804, and iron framing soon became the norm. With a rigid iron framework in place, a building no longer needed thick, load-bearing masonry walls; they could be replaced by thin screens of bricks or omitted altogether. Engineers could also replace a traditional wooden roof truss with one made from wrought iron. Once again, railway engineers were at the forefront of design innovation, and Robert Stephenson combined the structural benefits of cast-iron pillars and a wrought-iron roof frame to produce the first of those most iconic structures of the Victorian age, the train shed. The brand-new Euston Square station, where visitors to the Great Exhibition from Edinburgh, Dublin, Newcastle, Manchester, and Birmingham alighted, was a huge open area, with its long curved roof trusses held aloft by widely spaced pillars that rose from the platforms.

But even the Euston Square station was overshadowed by the immensity and grace of the Crystal Palace itself, whose design had evolved from another innovative type of building: the glasshouses that were being raised to house the botanical spoils of Britain's expanding empire. Britain's aristocracy had for centuries built orangeries and hothouses where they could grow exotic fruit, but these buildings were limited in size and had to face south to enable their small windows to let in enough light. The breakthrough in hothouse architecture was made by the Duke of Devonshire's head gardener, Joseph Paxton, who in 1841 designed and built the "great conservatory" to house the duke's Amazonian water lilies on the grounds of his Chatsworth home. The great wrought-iron-and-glass roof of this structure was stiffened not by a truss, but by being corrugated, like pleated palm fronds, so that it supported itself and the glass panes in between. Other large glasshouses soon followed, notably Decimus Burton's 1844 Palm House at Kew Gardens. So when the competition for the design of a building to house the Great Exhibition failed to come up with a satisfactory conventional solution, the committee were quick to accept the radical design suggested by Paxton.

In what was basically an upscaling of a glasshouse, Paxton developed a building that also incorporated many features that made it practicable and which were to become common aspects of industrial design. The corrugated roof was devised so that it could be constructed as it went along; glaziers traveled along it in carts, whose wheels ran in grooves along the structure. The grooves were also intrinsic aspects of roof drainage, as they channeled rainwater to the hollow pillars down which it could run to the ground. But the most important aspect of the design was that the structure was prefabricated and made from just a few standard designs of iron castings that could be made in bulk in a factory, shipped to the capital, and slotted together on-site. The whole structure took only eight months from the design phase to completion, the ironwork and glass panes being manufactured in Smethwick, Birmingham, taken by train to London, and erected in five months by a team of five thousand navvies. When the exhibition was over, it was taken down just as quickly, sold and reassembled in Sydenham, southeast London.

Inside the Crystal Palace, the range of attractions on display was vast and eclectic. The exhibits were complemented by a central glass fountain surrounded by palms and tree ferns, and two giant Egyptian statues. Many of the exhibits, especially those from Britain's empire, were exotic items: raw materials and examples of the indigenous cultures of the colonies' peoples. There were costumes, decorations, and traditional weapons from places as far afield as New Zealand and Canada; and huge diamonds, silks, and elephant trappings from India.

There were displays of decorative and fine arts from all around the world: fabrics, ceramics, statues, and paintings, showing what humans could produce without the aid of machinery.

But most of all, the exhibition showcased British industry. There were a myriad of machine-made artifacts, ornaments, and knick-knacks that British factories could now churn out quickly and cheaply. And there were the ingenious machines and powerful engines that produced them. There were spinning machines, looms, stationary and marine steam engines, railway locomotives, a steam hammer, a riveting machine, a hydraulic press, one of the hydraulic rams used to raise the Britannia Bridge, and a bewildering variety of machine tools. Many of the machines were even shown in motion, powered by a boiler set up outside the main building. The display captivated and overwhelmed even Queen Victoria, a person not generally known for her engineering interests. All this amply justified the organizers of the exhibition in their desire to demonstrate the dominance of British technology and reaffirm its place as the workshop of the world.

However, there were also indications that all of this was soon to change: the exhibition also housed inventions from around the world that would soon start to topple Britain as the world's leading industrial nation. From France and Belgium there were fast-rotating water turbines that could extract power more efficiently from rivers and streams than huge clumsy waterwheels. From Germany, the heavy-engineering company Krupp showed a giant ingot of the latest new wonder material, steel, along with their seamless-steel wheel rims for steam locomotives, and a steel canon that showcased Germany's growing military might. Another of their exhibits pointed even more clearly to the future: the electric telegraph, recently developed by another future engineering giant, Siemens. The American section also signaled the rise of the country's agriculture and industry. The huge potential of America's Great Plains, following the invention of John Deere's steel plow, was demonstrated by Cyrus McCormick's reaping machine. And creating even greater sensations were the guns of the rifle makers Samuel Robbins and Richard Lawrence, and Samuel Colt's newly invented revolver pistols. They demonstrated the American lead in machine tooling, as the precision of their engineering tools had enabled these manufacturers to produce totally interchangeable parts for their weapons. This was to be the key to the future of mass manufacturing and to the development of the assembly line, which American industry was to pioneer.

Overall, though, the exhibition was a triumph for Britain, not least because of its marketing. It would remain in everyone's mind the Great Exhibition, while the beauty of the Crystal Palace kept intact the Victorians' vision of Britain as a rural idyll. Artists were happy to paint it, even while they ignored the actual machinery and the other aspects of industrialization. Only William Turner had dared depict the rising power of steam in such paintings as *Rain, Steam and Speed* and *The Fighting Temeraire*.

But though artists ignored the technology that built the nation, the adjective *great* did sum up most early Victorians' perception of their country and their attitude to the achievements of its engineers. This attitude was embodied most obviously in the creations of Isambard Kingdom Brunel, which included the Great Western Railway; and his three ships, the *Great Western*, *Great Britain*, and *Great Eastern*. And by mid-century the mechanical power that Britain could produce really was great. For the first time its steam engines could produce more power than its workforce could develop manually; the installed power of steam engines in 1850 was 1.5–2 million horsepower (1.1–1.5 gigawatts), whereas its 14 million strong workforce could only have produced around 1 million horsepower (0.75 gigawatts). And the machines could perform hyperhuman feats. The largest engines could deliver over five hundred horsepower (370 kilowatts), each one replacing the labor of five thousand men. The largest steam hammers could produce a blow from their cylinders of five megajoules, fifteen thousand times what even the strongest navvy could deliver with a sledgehammer. And the largest hydraulic presses could deliver a crushing force of over fourteen thousand tons, fifty thousand times what a man can bench-press.

The steam-powered machinery transformed the productivity and output of British industry. For instance, British cotton production rose twelvefold, from 23,000 tons a year in 1800 to 280,000 tons in 1850. But to achieve this, Britain also needed to generate many times as much power. By 1850 it was burning 75 million tons of coal a year, five times as much as in 1800, producing seventeen hundred petajoules of energy, twenty-five times the amount its people obtained from the crops they grew in the countryside. Humans were now becoming far greater consumers of power than producers. And they were using up stores of energy that had been captured by plants from sunlight over millions of years.

Chapter 16

TRANSMITTING POWER

Today, the most obvious legacies of the Industrial Revolution are the railways, with their steam locomotives and wrought-iron bridges, but at the time the aspect of Britain that impressed and shocked visitors the most was the explosive growth and shambolic state of its cities. London, the center of the British Empire, was the largest, and fastest-growing, but the early nineteenth century also saw the proliferation of a new sort of urban settlement: manufacturing towns that specialized in a single trade. Within the textile industry the center for cotton was Manchester (also known as Cottonopolis); the wool trade was centered around the Yorkshire cities of Leeds and Bradford; the silk trade was based in the Cheshire town of Macclesfield; the linen trade around the Irish city of Belfast (or Linenopolis); and the jute trade around the Scottish city of Dundee. Around those centers, small towns grew into larger ones that specialized in even narrower aspects of production. Around Manchester, Oldham and Rochdale spun thread, while Bolton specialized in weaving plain cloth, and Preston produced fancier goods. And the port town of Lancaster made linoleum flooring by pressing imported linseed oil into rough cotton cloth. Similar towns and cities soon started to grow across the rest of Europe and in the fast-growing USA.

Within these towns, entrepreneurs built vast steam-powered mills where most people worked, but these were not the only impressive buildings. Local governments were quick to erect ever more grandiose town halls, law courts, prisons, museums, trading exchanges, covered markets, and churches to show off the prestige of their town. But the city fathers and mill owners alike paid little thought to how they would house the huge numbers of workers who were pouring into the new urban areas to take up jobs in the factories. The densely packed terraced houses and tenements that speculative builders threw up rapidly became overcrowded and transformed into foul-smelling, cold, dark and crime-ridden slums, prone to epidemic diseases such as smallpox, typhus, scarlet fever, measles, and cholera.

So beneath the surface gloss, urban life was far from great, particularly for the workers. Though they may have been paid slightly better wages than agricultural laborers, their shift hours were long and working conditions were harsh and dangerous. Whole families, men, women, and children, had to labor to make ends meet, and periodic fluctuations in trade often saw them laid off. And since, unlike the workers in rural mills, they did not have gardens, they had no chance to grow their own food; when unemployed or on strike to improve their pay, many people starved to death, most notably in the "hungry forties." It was not surprising, therefore, that life expectancy in the largest cities such as London and Manchester actually fell in the first half of the nineteenth century.

All this did not go unnoticed. Novelists and journalists, such as Charles Dickens in London, Victor Hugo in Paris, and Elizabeth Gaskell in Manchester, railed against the dreadful conditions of the poor. The son of a Manchester factory owner Friedrich Engels documented the problems in his 1845 polemic, *The Condition of the Working Class in England*. The work provided part of the inspiration for his joint publication in 1848 with Karl Marx of *The Communist Manifesto*. Combined with the wave of revolutions that spread across Europe in the same year, it seemed that capitalism's exploitation of the proletariat would inevitably lead it to be overthrown by workers' revolts and replaced by communism. Despite middle-class flight away from cities to the growing suburbs, politicians finally had to admit that the problems created by unfettered industrialization had to be addressed—especially since they even affected their own lives. London's Great Stink in the hot summer of 1858, for instance, made working conditions in the riverside Houses of Parliament unbearable because of the stench from the sewage-filled Thames.

But it is all very well to campaign or legislate against social ills; you also need to find practical ways to relieve them. It was one of the greatest triumphs of nineteenth-century engineers that they came up with technological solutions to problems that had once seemed intractable. They managed the task by developing new ways to transmit the power obtained from coal and water all around urban areas, to make them pleasanter and more convenient places in which to live and work.

One of the first problems that engineers tackled was the lack of public lighting, which was one of the causes of high crime rates. Traditionally, people had always lit their own homes using the feeble flickering flames of candles; the wealthy used expensive beeswax, while the poor had to make do with foul-smelling tallow, or animal fat. By the eighteenth century, candles had largely been replaced by brighter, more reliable, oil lamps, powered by whale oil. However, because of the long and perilous supply chain—whalers had to sail halfway around the world and brave untold dangers to obtain the oil—it was always expensive, so using it to light the streets was not economical. They remained dark, dangerous places at night and were hard to navigate, particularly in the regular smogs that blanketed them. In the end, engineers turned to the fuel that had caused most of the smogs—coal—to provide the solution to lighting them.

They could not use coal itself, since when it is burned in small quantities, it merely smolders with a dull red glow. Instead, they found that by heating certain forms of coal in the absence of air, they could produce a flammable gas. Iron smelters had been heating bituminous coal for over a century to produce coke. However, if people used cannel coals instead, shales that were particularly rich in hydrocarbons, the process also produced a mixture of methane and carbon monoxide, along with small carbon particles, a mixture that burned with a bright yellow flame. Engineers could collect the "coal gas" above water in large circular gasometers—the huge drum-shaped buildings that used to grace our cities—and pump it through pipes to consumers. Coal gas was far cheaper than whale oil and, when burned, produced a brighter, more even light; so by the beginning of the nineteenth century it was already being pressed into service in textile mills, whose owners were keen to extend their working hours through the night. Wealthy people also started to light their homes with the gas, and the first public gaslights were set up in Pall Mall, London, in 1807. The first commercial gas plant was the London and Westminster Gas Light and Coke Company, whose gas lit Great Peter Street by 1812 and Westminster Bridge by 1813.

Large numbers of small gasworks were soon set up all round the world, and by 1850 most cities and towns had gasworks, which lit not only the streets and factories but people's homes. The tar, an important by-product of gas manufacture, started to be used for a wide range of purposes. Most notably it became the feedstock of new chemical industries that produced artificial dyes such as the world's first, William Perkin's mauve. Engineers also started to investigate how they could exploit one of the potential drawbacks of gas—its potential to explode—to produce mechanical power. The first gas engines developed in 1860 by the Belgian engineer Étienne Lenoir were the gas equivalents of Newcomen's early steam engines. They burned gas at low pressures, moving the piston within a cylinder to produce a partial vacuum, so that the power stroke was driven by atmospheric pressure. However, since they operated at low temperatures and pressures, Lenoir's engines were heavy and inefficient.

The real breakthrough to make a useful engine was made by the

German engineer Nicolaus August Otto, who compressed a mixture of gas and air in the cylinder before igniting it. Otto's four-stroke gas engines were the forerunner of all future internal combustion engines, as we shall see in the next chapter. They were much quicker and easier to start than steam engines and so could be turned on and off whenever they were needed. Working at higher pressures and temperatures than Lenoir's engines, they were also far lighter and more efficient; the largest machines, weighing only half a ton (450 kilograms), could generate 30 horsepower (22 kilowatts), giving them a power-to-weight ratio around ten times higher than a steam engine. Crossley Brothers of Manchester bought the worldwide license to produce Otto gas engines and built over thirty thousand by the end of the century; businesses used them to power all sorts of small workshops.

By the 1880s, gas had proven its worth in distributing both light and power across cities, though the feebleness of the light it produced and the low power of gas engines hindered further progress. Toward the end of the century, though, engineers overcame some of its drawbacks. In 1891, the invention of the gas mantle by the Austrian chemist Carl Auer von Welsbach enabled gas to produce a much more brilliant, white light, while gas companies developed and promoted gas stoves and gas fires. But by then gas faced competition from two other methods of power transmission, which in some respects were even better: hydraulics and electricity.

The unlikely idea of using water to transmit power was born out of public schemes to supply water to the growing cities, schemes that were to end the worst scourge of urban life, waterborne diseases. From the beginning of the nineteenth century, urban corporations sought to obtain clean water from surrounding areas and pump it to the streets and into the houses of the people. In most of the towns in the north of Britain, the task could be accomplished without expending much energy.

Civil engineers such as John Frederick Bateman built huge dams in the surrounding hills to create gigantic reservoirs and channeled the water by gravity down to the city through cast-iron pipes. Manchester, for instance, was supplied by water from the Longdendale reservoirs in the nearby Peak District, and later from Thirlmere in the Lake District, ninety miles away. Liverpool built Lake Vyrnwy in North Wales, while in the USA, New York City built a series of reservoirs in the Croton River catchment to supply a population that had increased threefold, from sixty thousand in 1800 to two hundred thousand in 1830.

The situation was rather different in London, which had no surrounding hills from which to collect rainwater. Engineers had to pump clean water from the Thames, twenty miles upstream from the capital, into raised storage reservoirs, from where, via filter beds, it had to be pumped by gigantic steam engines up into towers to create a constant head of water and pipe it eastward into London. This created the bizarre new landscape around the outer suburbs of southwest London where I grew up, a place where the roads and towns are surrounded by what look like flat-topped hills, and where the banks of the Thames are flanked by huge rectangular settling basins and gigantic Victorian engine houses.

In 1845, a Newcastle solicitor, William Armstrong, realized that the pressure of water in mains could also be used to power machinery. Armstrong was fortunate that Newcastle was a perfect test bed for his ideas. The docks along the River Tyne lay some two hundred feet (61 meters) below the main town at the foot of the steep Tyne valley, so the mains' water pressure was high, around eighty-five pounds per square inch (0.6 megapascals). This was around twice the pressure inside the cylinders of contemporary steam engines, so the water contained a lot of potential energy. Armstrong designed a crane whose pistons were driven not by steam but by water. The crane proved so successful that he soon received orders for cranes not only from Newcastle but also from ports all around the country.

Port authorities soon realized that high-pressure water would be the ideal way to transmit power around their docks to drive all the machines that they only needed to run intermittently. Rather than having lots of small, inefficient steam engines to power each of their machines, they could link up their cranes, dock gates, capstans, and riveters to a high-pressure water main. A single large steam engine could power a pump that provided the high-pressure water for all of their machines, machines that, unlike steam engines, could be switched on and off at will. The only other element that such a system needed was a device to maintain a constant water pressure and to store energy for peak power demands. In the first major port that was powered by water, Grimsby on England's east coast, Armstrong achieved both of these aims by pumping water two hundred feet (61 meters) up to a thirty-six-thousand-gallon (140,000-liter) tank that was housed in a gigantic brick tower. This hydraulic accumulator, as it was known, provided the same water pressure as Newcastle's mains. Grimsby Dock Tower, which still stands, is a glorious structure. Based on the design of the Torre del Mangia next to the Palazzo Pubblico in Siena, it is over three hundred feet high and is visible for miles across the Humber estuary.

Even before the Grimsby Dock Tower was finished in 1852, however, it was shown to be something of a white elephant. As Armstrong was building another hydraulic system for the nearby port of New Holland, he realized that the ground there was too soft to support such a high tower, so he came up with the idea of using a heavy weight instead of tall column of water to supply the pressure. Using such a weight-loaded accumulator, he could supply the pressure using a much shorter, cheaper building, and he could also develop much higher pressures, so the pistons in his machinery could be much smaller, they could be more powerful, and they would use far less water. All subsequent hydraulic systems used weight-, spring-, or compressed-air-loaded accumulators, including the one in the short crenelated accumulator building that replaced the Grimsby Dock Tower in 1894. It was just fifty feet tall and provided the water pressure using a three-hundred-ton weight.

Over the 1860s and 1870s, ports all around Britain and the Atlantic coast of Europe were fitted with hydraulic machinery, and large numbers of hydraulically powered swing bridges were also built. The first was Armstrong's Low Level Bridge over the Tyne just downstream from Robert Stephenson's High Level Bridge. It enabled Armstrong to access his Elswick works via both road and water. Most hydraulic swing bridges, however, took railway lines over rivers and canals, while the most bizarre, the Barton Swing Aqueduct, took the old Bridgewater Canal over the new Manchester Ship Canal. Still in use today, the two ends of the aqueduct first have to be closed off with gates to hold the water in before it swings around to allow ships to pass. The most spectacular of all the hydraulic structures are the huge boat lifts at Anderton in Cheshire, and the even larger one at La Louvière in Belgium. They raise and lower canalboats and barges from one canal to another using massive hydraulic rams.

In the 1870s, the British engineer Edward Ellington came up with the idea of setting up high-pressure mains-water systems to distribute mechanical power not just around ports but around whole cities. The first was built near where I live in Hull, but Glasgow, Manchester, and Birmingham soon followed in Britain, and systems were set up as far afield as Antwerp, Melbourne, Sydney, and Buenos Aires. The largest of all the systems was the one serving London, which at its peak had 180 miles (290 kilometers) of high-pressure mains water maintained at seven hundred pounds per square inch (4.8 megapascals), and driven by steam engines that provided an average power of fifteen hundred horsepower (1.1 megawatts). It not only powered London's dock machinery but workshops, hotel lifts, the safety curtains of theaters, and most spectacularly of all the Savoy Hotel's moving cabaret stage and the rising floor of the Earls Court Exhibition Centre. Though now almost totally forgotten, hydraulic mains worked profitably for the best part of a century. London's, for instance, was in operation until well into the 1970s. The systems were finally rendered uncompetitive due to damage by German bombing in World War II and corrosion of the pipes.

Fortunately, however, many of the large hydraulic structures remain, though now converted from steam to electric power, and many of them retain their hydraulic accumulators. You can still marvel at their machinery and admire their operation, none being more spectacular than that most iconic structure, Tower Bridge in London. The two lifting arms of this huge bascule bridge, completed in 1894, were powered by two 360-horsepower steam engines, which pumped water into two weight-loaded hydraulic accumulators at the foot of each tower. When a ship approached, the seven-hundred-pounds-per-square-inch (4.8-megapascal) water pressure lifted the two 1,070 ton arms to let it pass. The system, backed up by London's hydraulic mains in case of the failure of its engines, worked without incident until 1976, when the steam engines were replaced by electric motors. Fortunately, however, you can still see the whole system, including its original steam engines, as the bridge is now open as a tourist attraction.

An even more surprising tourist attraction commemorates this first age of hydraulics: Geneva's famous Jet d'Eau water spout. It is a relic of Geneva's own hydraulic system, which dates back to the 1880s. The system, which supplied the many small workshops of this watchmaking city, was powered not by steam, but by a Jonvil water turbine located where the River Rhône exits Lake Geneva. The original spout, which reached a height of one hundred feet (30 meters), was set off at the end of each working day to relieve the pressure of the accumulator when the workmen shut down their tools. The sight proved so popular to visitors of this attractive but rather dull city that, when the hydraulic system shut down, the authorities re-created the spout, positioning it farther into the lake and making it a much more spectacular 460 feet (140 meters) high.

What doomed hydraulic power systems was not their inefficiency, but two inherent disadvantages. First, they were always vulnerable to frost, which might explain why they were never used along the east coast of America, where the winters are long and cold. Second, they could only supply mechanical power, not light. So in the end, hydraulic systems lost out to the first technology that was based on scientific discoveries rather than just basic mechanics, a technology that could provide not only mechanical power but also light and heat—electricity.

The first battery that produced a continual flow of electricity was the electric pile, invented in 1800 by the Italian physicist Alessandro Volta. Batteries were gradually improved over the first half of the nineteenth century, the first long-lasting stable cell being the Daniell cell. Invented by the British chemist John Frederic Daniell, this was the first battery to generate useful amounts of power. Daniell cells were used to provide the current for the electrical telegraphs that were developed from the 1830s onward, and which initiated the age of instant long-distance communication. Such batteries were far too expensive and heavy to provide useful amounts of light or motive power, but fortunately by the 1820s scientists were showing how electricity could be produced by motion, and in turn how electricity could power motion.

In 1820, the Danish physicist Hans Christian Ørsted showed that passing an electric current through a wire produces a magnetic field, demonstrating for the first time the connection between the two phenomena. Soon after, in 1821, the British scientist Michael Faraday showed that he could convert electricity into motion by running a current through a piece of wire that was set in a magnetic field; the wire rotated around the magnet. In 1831, Faraday also showed the reverse: that he could convert motion into electricity simply by rotating a conducting plate through a magnetic field. The concepts of the electric motor and the dynamo had been born, though it took decades for these concepts to be converted into practical machines. The first workable DC electric motor and first dynamo using an electromagnet were both designed by the Hungarian Anyos Jedlik, though the first commercially successful models of these machines were only developed in the 1860s by engineers such as Charles Wheatstone, Werner von Siemens, and Samuel Varley.

The basic technology to produce electrical energy and to use it to provide mechanical power had been solved, but it proved more difficult to find practical ways of converting electricity into light. One basic technique: creating a continuous spark between two high-voltage electrodes separated by air or another gas had been demonstrated at the beginning of the nineteenth century by the British scientist Humphry Davy. However, the arc light, as it became known, was perfected only in the 1870s. Its power was also something of a limitation; though it produced a clear white light, it was two hundred times more powerful than a typical light bulb of today and so was only suitable to light streets or large buildings such as theaters. Early examples were used to light the Public Square in Cleveland, Ohio, in 1879. It was 1880 before Thomas Edison in America and Joseph Swann in Britain developed practical incandescent bulbs, in which electricity heated thin carbon filaments within vacuum tubes so that they glowed white-hot.

With two killer applications, to provide light and mechanical power, electrical transmission of energy suddenly became a commercial proposition, especially as all that engineers needed to transport the electricity were thin copper wires, far easier to lay than water or gas pipes and no trouble to maintain. Entrepreneurs set up local electric power stations as early as 1882: the first being in Holborn, London; the second in Pearl Street, New York. However, these direct-current (DC) power stations had a drawback. Since they operated at low voltages to keep them safe, the power lines had to transport large currents and so suffered from high transmission losses. Consequently they could only serve small areas. The obvious answer to reduce power losses was to produce alternating-current (AC) electricity, since engineers could raise and lower its voltage using transformers; they could transmit the electricity over long distances at high voltage before reducing the voltage nearer the homes where it was used. Once the Serbian American engineer Nikola Tesla had developed a practical AC electric motor in 1887, the argument for AC became unanswerable, and large AC central power stations were soon set up, the first being Sebastian de Ferranti's Deptford power station in East London in 1889. In the USA, George Westinghouse's AC system gradually won out in the "war of the currents" with Thomas Edison's DC system. Westinghouse famously won the bid to light the 1893 World's Columbian Exposition, which showcased the growing hegemony of American industry, just as the Great Exhibition had celebrated British achievements forty years before.

But even the new AC power stations suffered from another problem. The power they could produce was limited by the large size and low power of the steam engines for their generators, so they struggled to make a profit. The Deptford power station needed two of the largest steam engines available, powered by no fewer than twenty-four boilers, and even then it could provide no more than three thousand horsepower (2.2 megawatts). The problem was the slow movement of the reciprocating pistons and the huge vibrations they set up. The answer was to replace these reciprocating engines with a new breed of rapidly spinning generators. By the 1850s French and American engineers had perfected a series of small rapidly rotating water turbines that could extract energy from water much more cheaply and more efficiently than eighteenth-century waterwheels. Francis turbines helped extend industry into the uplands of the American northeast, while Pelton wheels were developed from the primitive hurdy-gurdy wheels used by prospectors in the Californian gold and silver rushes of the 1850s and 1860s. But most sources of hydraulic energy were located far away from habitation and industry, so hydraulic energy was otherwise ignored.

This all changed with the advent of AC electric generators, and engineers rushed to exploit free hydroelectric power. The first person to use hydroelectricity to light a house, in 1890, was William Armstrong, who powered his new country residence, Cragside, using one of James Thompson's vortex turbines, which was housed at the end of Armstrong's lake. Large-scale hydroelectric plants soon followed, the first major station being the Edward Dean Adams Power Plant, which was set up in 1895 at Niagara Falls, New York State. It generated an unprecedented fifty thousand horsepower (37 megawatts) of electricity, which was transmitted to nearby cities such as Buffalo. A whole series of hydroelectric plants soon followed in the mountains of northern Italy, which drove the industrialization of Italy's new economic powerhouses, Milan, Turin, and Genoa.

Few other cities or industrial areas were located anywhere near high mountain ranges that could produce appreciable amounts of power, so fortunately in the 1880s another large power source was being developed, one that combined steam with the rapid rotation of water turbines—steam turbines. The reaction steam turbines developed by the British engineer Charles Parsons used high-pressure steam to drive a series of multibladed fans, like those on a Wild West wind pump. The steam moves along the axis of the turbine hitting each set of blades in turn and causing them to rotate around their single axle, while a series of stator blades in between the moving blades keeps the steam moving in the axial direction. As it travels down the device, the steam loses pressure and expands, so the blades are made larger and larger farther along the axis of the turbine, allowing it to extract more of the energy in the steam.

Parsons turbines are more powerful than a steam engine of the same size and do not produce unwanted vibration. They also had the advantage that since they rotate at speeds of around three thousand revolutions per minute, they could power AC generators operating at fifty or sixty hertz without needing any gearing. By 1899 fifteen-hundred-horsepower (1.1-megawatt) turbines were being used in power stations, but this was just the beginning of their development.

By 1910 the largest steam turbines could produce thirty-five thousand horsepower (26 megawatts) with an efficiency of around 12 percent, equivalent to the mechanical work of four hundred thousand people, and were well on the way to becoming the powerhouses of the modern world. Shipbuilders started to use them to power their vessels, where their high power and efficiency and vibration-free operation enabled battleships and passenger liners such as the *Titanic* to grow ever larger and sail ever faster.

The benefits of an electric supply were not restricted to lighting streets and houses. Electrical gadgets of all sorts soon proliferated, powered by myriads of small, efficient electric motors. In factories each machine could be powered by its own motor, so they could be arranged into efficient assembly lines. And in homes they could be put to work to power all sorts of laborsaving devices, from vacuum cleaners to washing machines. But the greatest effect that the electricity supply had on the turn-of-the-century city was on transportation. As cities had got bigger and bigger, people had found it harder and harder to travel between home and work and obtain the goods they needed. Wealthy people could use their horse-drawn carriages or cabs, while the poor could use buses or trams pulled by overloaded nags, but the huge number of horses demanded a whole infrastructure of mews and stables. Horses, moreover, created vast amounts of waste that had to be cleared away by road sweepers and piled into huge "dust heaps" or transported out to market gardens. Toward the end of the nineteenth century, city authorities were looking for better ways of conveying people quickly and cheaply. Steam trams were heavy and polluting, while the steam locomotives used on London's first underground railway lines produced an unbearable choking atmosphere. New power systems were clearly needed to help transport people, just as they had already supplied industry and people's homes.

Many North American cities adopted cable cars, hitching the vehicles to cables that ran beneath the tracks and which were directed under the streets by complex systems of pulley wheels. Cable cars had the advantage that they could run even in hilly cities such as San Francisco, but the high friction in the cable system wasted over 90 percent of the energy supplied by the stationary steam engines that powered them, so most were soon shut down. In Europe, most trams were instead converted to electricity; they were supplied with power using a pantograph on the top of the vehicle, which contacted power lines overhead. In London, the underground railway was also quickly converted to electric traction, the trains connecting to the power supply using a third rail. Electric traction made the fully underground "tube" system practical. The first tube line was set up in 1890 running trains in tunnels from South London under the Thames, to the City of London and Finsbury to the north. The system is still in operation, acting as the eastern branch of the Northern Line. Electricity even enabled the development of faster verbal communication. By the 1880s houses and businesses were being connected together through telephone exchanges, so people could talk to each other without even needing to travel.

By the end of the nineteenth century, therefore, engineering projects had transformed cities. They were cleaner, lighter, safer, and healthier places in which to live. The growing middle classes could afford to move out, to live in pleasant suburban homes, with a growing number of laborsaving devices, and could get to and from work using public transport. This allowed cities to expand ever faster to coat the surrounding countryside with suburban villas. But like previous developments, this came at a cost. By 1900, Britain was using 225 million tons of coal a year, three times the amount it burned in 1850. And it was not just Britain whose energy needs were expanding. Worldwide energy demand had doubled from 1850, and the energy derived from coal was overtaking that from the traditional source of energy, wood.

In cities the changeover was more pronounced. Ninety-five percent of the energy used in nineteenth-century towns came from burning coal, whether to provide heat in open fires and stoves, in gasworks, or to drive steam engines that powered mills, factories, or electric power stations. This released unprecedented amounts of smoke, which combined with fogs to produce the choking smogs that could kill thousands of people, and for which cities such as London and Manchester were notorious. The smoke also coated the buildings with soot, creating Blake's "dark Satanic Mills," and giving the impression, until they were finally cleaned in the 1960s and 1970s, that the Victorians were a miserable bunch who only built in black stone. Artists avoided industrial cities like the plague, just as they had ignored the engineering marvels of the previous century, preferring to depict romantic rural idylls, or ludicrous medieval legends. Not until long after their heyday did a Manchester artist, L. S. Lowry, depict the inhuman scale of mill buildings and the alienating effect of the urban landscape around his home and was consequently ignored and patronized by the London art establishment.

Chapter 17

PORTABLE POWER

If you were transported back in time to the end of the nineteenth century, you might be impressed by the apparent modernity of the cities. The people would be moving to and fro along well-lit streets using electrically powered trams, or under them in electric trains just as we do today, and their houses would be filling up with laborsaving appliances. The countryside would be much more of a shock, however. You could have traveled to most towns and villages along the extensive rail network, but in between them you would have had to travel and move your goods in horse-drawn carriages or wagons or just walk along the rutted unmetalled roads. There would have been no machines visible at all in the fields, building sites, or quarries, and even within rural workshops, blacksmiths and carpenters would have been using only hand tools. The houses would have no mains water, electricity, or gas and would instead be supplied from nearby springs or wells, be heated by coal or wood fires, and be lit by candles.

Much of the lack of machinery stemmed from the limitations of coal-fired steam engines. For a start, they were unsuitable for transporting people around the country. Because they use steam as the working fluid, they need a heavy boiler as well as the cylinders where the mechanical power is actually generated, so they have to carry about large amounts of water. And since the steam has to be moved into the cylinders from the outside at each stroke, they are also slow-moving, which limits the power they can produce. Consequently, steam locomotives were too heavy to be used on unmetalled country roads, let alone on soft soil. Late in the nineteenth century, engineers did finally succeed in producing steam engines that could run along roads, but these "traction engines" had to be fitted with wide iron wheels to distribute their weight, and they were slow, lumbering machines. Some used the weight to their advantage, acting as steamrollers to flatten the surfaces of new roads, but most were used essentially as self-propelled stationary engines. Contractors could drive them from farm to farm, where they could link them up via an arrangement of belts to drive threshing machines. Showmen converted others to power touring carousels or fairground organs. The first steam-powered automobiles, meanwhile, were not only heavy, but suffered from the disadvantage of all steam engines, that they had to be fired up long before they were used. Which is why the drivers employed by the wealthy owners of cars are still known as chauffeurs, which is French for "stokers."

Ultimately, the modernization of the countryside was to rely on exploiting totally new sources of power, liquid hydrocarbons derived from mineral oil. But it took engineers decades to learn how to use them to generate mechanical power, to give humankind the freedom of travel not only around the country but across the seas and in the air. First of all, they had to learn how to burn them to generate light and heat.

People had been exploiting the mineral oils that seep out of the earth in the Middle East from the Bronze Age onward. The Sumerians had used them as a source of bitumen to waterproof their ships, and bitumen also helped the Babylonians build the walls of their great city. The Chinese were the first to use oil for fuel and to actively drill for it, and by the Middle Ages the Arabs were extracting oil and distilling it into flammable products for military purposes. However, oil was largely absent from Europe. Only in the middle of the nineteenth century did industrialists and chemists start to realize that they could extract oil from certain types of coal and shale simply by heating it up in a retort. The first shale oil was mined in Europe in Autun, France, in 1837, and soon afterward the Scottish chemist James "Paraffin" Young, developed distillation techniques to separate shale oil into several useful fractions: light paraffin oil or kerosene; heaver lubrication oil; bitumen or asphalt; and paraffin wax. In 1851 he set up the world's first major oil refinery in Bathgate, twenty miles west of Edinburgh, which stimulated the development of shale oil mines and refineries all across the area. Nothing now remains of this once-great industry except for the huge heaps of waste rock or bings that still tower over the flat countryside, but soon shale oil industries sprang up across Europe and the Americas.

The first major use for the new products of the oil industry was for lighting. Paraffin wax, once hardened using stearic acid, proved to be ideal for making candles, as it burns cleanly with a bright yellow flame and with a sweet smell. It was far cheaper than the alternatives: beeswax or even tallow. The liquid fraction, paraffin, or kerosene as it is known in the United States, meanwhile, burned with an even brighter yellow flame and was soon being used in preference to other oils. The Polish chemist Ignacy Łukasiewicz invented the first kerosene lamp in 1853. It had a wick, like all oil lamps, to draw the liquid up into the flame and allow it to evaporate so that it would burn, but in addition Łukasiewicz surrounded the wick and flame with a glass chimney to ensure a rapid upward draft of air past the wick. This created a flame that could be ten or more times as bright as that of a candle. With the flame protected from the outside, and the lamp itself cooled by the air flow, kerosene lamps were safer than traditional oil lamps and rapidly became the lighting source of choice for travelers and country dwellers alike.

Following the drilling in 1859 of the first true oil well in the United States by Edwin Drake, and the subsequent Pennsylvanian oil rush of the 1860s, the price of oil fell below that of shale oil. The oil was easier to extract as it could be pumped or simply allowed to gush out of the ground and required no heat treatment. Consequently the kerosene it contained became even cheaper, and inventors rushed to develop a range of kerosene heaters and kerosene stoves that could replace coal-fired burners in the countryside. None of them were totally satisfactory as they relied on the fuel's evaporating from wicks, just like candles and kerosene lamps. This resulted in incomplete combustion of the fuel, producing a smoky flame that quickly coated the apparatus with soot. Only in 1892 did the Swedish inventor Frans Wilhelm Lindqvist overcome these problems by developing the Primus stove. This works by forcing kerosene under pressure through a preheated burner head, which evaporates it, before it sprays out through a nozzle, mixes with air, and undergoes complete combustion, producing a hot blue flame. Primus stoves quickly became popular with adventurers and explorers, helping them reach the north and south poles in the early twentieth century and finally in 1953 to conquer the summit of Everest. I used my parents' old Primus stove on my 1970s camping expeditions and remember it fondly. The only downside was the clinging odor of paraffin that pervaded those holidays.

There are other ways of evaporating kerosene, though, rather than heating it. Modern oil-fired central-heating systems use cleverly designed nozzles that atomize the fluid into thousands of tiny droplets as they emerge, droplets that quickly evaporate and burn within the boiler to produce a strong clear flame. Today, kerosene boilers are used in millions of rural dwellings that are too isolated to be served by gas mains. And in the late nineteenth century engineers developed similar methods to enable them to use kerosene and other liquid fuels not merely for lighting and heating, but in a new breed of internal combustion engines that could power motor vehicles.

Given the impracticalities of steam power, and the weight and high cost of early batteries, which put paid to the practicality of electric cars, the obvious solution to powering early road vehicles was the internal combustion engine. Otto's four-stroke gas engines of the 1860s had shown that such machines could be powerful and efficient, but gas engines could not be the answer; the low energy density of the gas meant that the engines had a low power output for their weight, and transporting gas around under pressure was difficult and dangerous. Otto experimented with using explosive coal dust, but the soot particles it produced wore down the piston rings and seized up his engines in minutes.

Engineers turned instead to the only alternative: liquid hydrocarbon fuels. They had the advantage that at forty to forty-five megajoules per kilogram they had twice the energy density of coal or coal gas, and they could easily be transported. One potential fuel was the mixture of light hydrocarbons that were created as a by-product of making coal gas, but this had such a low boiling point and was so volatile that it frequently exploded prematurely, damaging the cylinders. A plentiful alternative was kerosene, but as this has a much-higher boiling point it was difficult to vaporize in the cylinder so that it would burn at all. Only in the 1880s did engineers overcome these problems. In the hot-bulb engines invented by the British inventor Herbert Akroyd Stuart and developed by Richard Hornsby, the fuel was vaporized in a red-hot chamber before being let into the cylinder. In the oil engine, invented by William Priestman, in contrast, the fuel was atomized as it was injected through a nozzle into the cylinder, as in a modern oil-fired central-heating boiler. Both sets of machines were reliable, but worked at rather low compression ratios and at low speeds, so were rather inefficient and low powered. They served well on farms, where they provided the power for threshing machines and other machinery, but they proved too heavy to be used in road transport.

The two most successful designs of internal combustion engines used hydrocarbons that were heavier than the residue of coal gas production but lighter than kerosene. Gasoline, or petrol, contains hydrocarbons with between four and twelve carbon atoms, while diesel contains hydrocarbons with between twelve and twenty. Gasoline proved the easiest fuel to use, and four-stroke engines were being developed in

Germany as early as the 1880s. The volatile fuel was mixed with air moving rapidly through a narrow tube in the carburetor, before being introduced into the cylinder, compressed, and ignited with a spark produced using an electric coil. The first gasoline-powered car was Karl Benz's Patent-Motorwagen of 1885, a tricycle that had a single-cylinder engine that produced 0.9 horsepower (662 watts) and weighed only 220 pounds (100 kilograms), so it had a power-to-weight ratio of almost seven horsepower per ton, several times that of a typical steam engine. In the same year, Gottlieb Daimler and Wilhelm Maybach produced the first gasoline-powered motorbike, the Reitwagen, propelled by a 0.5-horsepower (368-watt) engine weighing only around fifty kilograms (110 pounds).

The problem of using the heavier fuel, diesel, took longer to solve. Rudolf Diesel produced the first prototype engines in the early 1890s, and only in 1898 did he produce the first commercial diesel engines. Diesel engines work by compressing air in the cylinder so hard that it heats up enough to reach the combustion temperature of the diesel fuel, which is then injected directly into the engine. Diesel engines work at higher temperatures and pressures than gasoline engines, so are even more efficient, up to 50 percent for large, slow-moving engines, but this comes with the disadvantage that they have to be stronger and heavier. Consequently, the first diesel engines were stationary, used to replace steam engines. They were gradually joined in the early twentieth century by engines to power ships, trucks, and buses. Given their high efficiency, many diesel engines were also used as mini–power stations to generate electricity for farms, quarries, mines, and country houses. And German railway engineers used the same idea to develop the first diesel trains, in which the engine generated electricity, which in turn powered electric motors that moved the wheels. Finally in the 1930s, the introduction of the swirl chamber enabled diesel engines to run at high speeds, making diesel-powered automobiles more competitive.

At the start of the twentieth century the main factor restricting the use of internal combustion engines was the supply of the fuels. Both oil shale and crude oil are largely made up of long-chain hydrocarbons, so distilling them gave large quantities of kerosene, lubricating oils, and waxes but little gasoline or diesel. Consequently, in the first mass-production car, the Model T Ford of 1908, the engine was designed with a unique carburetor so that it would work not only with gasoline but also kerosene and even ethanol. Fortunately, chemical engineers in Russia and the USA developed novel ways of breaking up large hydrocarbon molecules into smaller ones by heating them up under pressure—a process known as thermal cracking. Large quantities of light fuel could at last be produced, and motor vehicles started to take over the roads from horses and carts, a process speeded up by two additional factors.

Bicycle engineers had already solved many of the difficulties in producing lightweight vehicles in which to mount the new lightweight engines. They had pioneered structures made from tubular steel frames and lightweight wire-tensioned wheels. They had developed ball bearings that enabled the wheels to run more smoothly. They had invented differential gears that allowed the back wheels of tricycles to rotate at different speeds from each other and so allow vehicles to run around corners. And they had developed pneumatic tires that gave a smoother ride. Motor vehicles also benefited from improvements in the roads; their gravel surfaces were starting to be consolidated with tar or bitumen to produce smoother asphalt and Tarmac road surfaces. The countryside became more accessible from the cities, while urban roads started to be clogged with cars and trucks, rather than horses and carts.

Even before internal combustion engines had revolutionized road transport, they had enabled people to conquer the final element—the air. Throughout the nineteenth century, would-be aviators had been investigating how they could build heavier-than-air flying machines. The British baronet Sir George Cayley had built and flew (with his coachman on board!) the first man-carrying glider in 1832 and experimented with model aircraft driven by propellers powered by rubber bands. The German Otto Lilienthal built and flew a series of gliders in the 1890s, investigating how he could use his weight to control the aircraft. However, early attempts at a machine that could undertake sustained powered flight, such as the Frenchman Clément Ader's *Éole*, were hindered by the low power-to-weight ratios of their steam engines. The key to the success of the Wright brothers in making the world's first powered flights at Kitty Hawk, North Carolina, in 1903, therefore, was the engineering of their homemade gasoline engine; it was capable of providing twelve horsepower (9 kilowatts) while weighing only 180 pounds (82 kilograms), a power-to-weight ratio of 110 watts/kilogram, over ten times higher than that of the engines used in the first car and motorbike. The brothers also benefited from their experience in making lightweight structures in their bicycle shop, and years of experiments investigating methods of controlling gliders and developing efficient propellers.

Throughout the first half of the twentieth century most of the improvement in the performance of aircraft was due to the development of more powerful engines. By World War I, aircraft engines such as Clerget rotaries could typically produce one hundred to two hundred horsepower (75–150 kilowatts) with power-to-weight ratios of around five hundred watts/kilogram, giving their planes maximum speeds of around 100 mph. By World War II, engines such as the Rolls-Royce Merlin could develop ten times the power, around fifteen hundred horsepower (1100 kilowatts), with a power-to-weight ratio of thirteen hundred watts/kilogram, By this time propeller-driven airplanes were reaching speeds of 400–500 mph (650–800 kph) but were nearing the limits of their performance; the propeller tips were approaching the speed of sound, at which they became ineffective. Engineers in Britain and Germany were therefore developing a new type of engine that combined the benefits of burning a liquid fuel, like internal combustion engines, with the rapid rotating blades of the steam turbine—gas turbines or jet engines.

Compared with piston engines, jet engines are far-simpler machines, having just a single moving part, the turbine. The fan at the front compresses air back into the combustion chamber, where it mixes with vaporized kerosene fuel and burns. This produces a fast rearward stream of hot expanded gas that impacts a smaller set of turbine blades that drive the engine around, before the gas exits at the back of the engine to provide thrust. Even before the end of the World War II, jet engines had power-to-weight ratios of over two thousand watts/kilogram and were propelling aircraft at over 500 mph (800 kph). The first commercial jet aircraft, the de Havilland Comet, was launched in 1952, carrying forty passengers at speeds of up to 450 mph (700 kph), while the first supersonic jet aircraft was the North American Super Sabre, which first flew in 1953.

By the middle of the twentieth century, engineers had harnessed the power from liquid hydrocarbon fuels to free people to travel wherever they wanted and at unprecedented speeds. Cars could speed along roads as fast as trains, shrinking distances and making the country as accessible as the town. Passenger liners, powered by steam turbines that ran on oil rather than coal, could take people halfway across the world without having to refuel, and diesel-powered cargo ships could move goods cheaply between continents in days. Aircraft could move people even faster, crossing continents in hours.

But the new engines speeded up death just as much as they speeded up life, since they enabled military engineers to develop new forms of mechanized warfare. On land, diesel-engine trucks transported troops and guns to and from the battlefield, and tanks, fitted with tracks, could roam across country at speeds of up to 30 mph (48 kph), bringing the firepower of cannon with them; the 75 mm gun of a Sherman tank, for instance, could fire its fourteen-pound (6.35-kilogram) shells at a speed of 1,400 mph (620 meters per second), giving them a range of almost two miles (3.2 kilometers) and transferring 1.3 megajoules of kinetic energy, and 3 megajoules due to the explosion of its charge, to its target.

At sea, battleships with oil-fired turbines could steam at 30 mph firing fifteen-inch shells that could fly eighteen miles (29 kilometers) and deliver three hundred megajoules of destruction. Below the sea, diesel engines charged the batteries of submarines that could fire 1.5-ton torpedoes that produced fourteen-gigajoule explosions that could sink the largest ships. And in the air, bombers traveling up to 300 mph could carry high-explosive bombs weighing up to ten tons that delivered twenty-five gigajoules of energy. Rocket engines could deliver warheads even farther and faster. Germany's V2 rocket was supersonic and had a range of over two hundred miles (320 kilometers). With two thousand pounds (907 kilograms) of explosives in its nose cone, it produced some four gigajoules of energy when it hit its target. Liquid-fueled engines enabled people to kill each other for the first time in their thousands, tens of thousands, and hundreds of thousands, even when they were civilians who lived hundreds of miles away. And by the 1950s airplanes and rockets could also transport the most deadly weapons of all time: atom bombs and hydrogen bombs. The uranium-powered atom bomb that was dropped on Hiroshima released some sixty-eight terajoules of energy, and the plutonium weapon dropped on Nagasaki released even more, some eighty-eight terajoules. The hydrogen bomb is twenty-five hundred times as powerful as the Nagasaki bomb, capable of releasing up to two hundred petajoules of energy, and so two thousand million million times more powerful than a single human combat soldier!

Powering all this movement and destruction used a lot of energy. By 1950, global energy use had doubled from 55,000 petajoules per year in 1900 to 110,000 petajoules, five times as high as in 1800, most of the rise being due to the explosive growth in the use of hydrocarbon fuels. And since then, new ways of using energy have enabled us to carry that trend forward, transforming the world of work and magnifying our impact on the world around us still further.

Chapter 18

THE HEGEMONY OF THE MACHINES

One of the greatest difficulties in understanding modern history is to get a proper perspective, to see the bigger picture, and to identify the major trends. We are too close to the events that we are living through, and the journalists and commentators who inform us about what is happening tend to focus on ephemeral cultural trends, the shifting alliances and machinations of politicians, and the power struggles and wars between ideological blocs. But while all these events matter, they scarcely affect the ways in which we live our lives. And when I look back at the sixty years through which I have lived in the developed West, the world around me looks superficially much the same as it always did.

Most of us still live in the same sort of houses that we lived in sixty years ago, houses that are still supplied in the same way with gas and electricity, filled with laborsaving electrical appliances and power tools. Most of us still drive cars that are powered by gasoline or diesel and travel along the same Tarmac-covered roads that we always did. Goods are transported in the same way overland, on electric- or diesel-powered trains, and in trucks and vans, while over sea they are moved in ships driven by diesel engines. People still travel across continents in subsonic aircraft powered by jet engines. And electricity is for the most part generated by steam, water, and gas turbines, with a small but growing percentage produced by wind turbines and solar panels.

227

Ostensibly, then, everything seems to be familiar and unaltered. Yet if we look beneath the surface, we can see that there has been a total revolution in the way that we interact with the world around us. Technological advances have enabled machines to transform work: lifting output, improving productivity, raising living standards, and spreading wealth across the globe. The past sixty years has seen the hegemony of machines and the eclipse of manual labor, so fewer and fewer people have jobs that rely on their using their muscles. And for once, artists produced works that actually showed people working, drawings and paintings that illustrated the life people led before the machine revolution and which help us to see how utterly things have changed.

During the desperate early days of the World War II, the UK government, eager to raise morale and foster a feeling of togetherness, paid large numbers of artists to go out into the world to document the war effort. Many of these "war artists" traveled with the armed forces, sketching soldiers in battle or relaxing behind the front lines. But many other artists concentrated on depicting the labor of the millions of men whose work was just as important for victory. They painted shipbuilders on the River Clyde, "Bevin Boys" hewing out coal, and navvies laying down runways. But since so many of the men had been called up into the forces, large numbers of the workers on the home front were women, who were for the first time ever painted wearing trousers or overalls and performing tasks other than nursing patients or spinning thread. The artists produced images of women in factories, machining the parts of guns or assembling aircraft; "land girls" on farms driving wagons, tossing hay, or sorting potatoes; and "lumberjills" in the forests, felling trees. These images show that even after 150 years of industrialization, society demanded more manual labor than ever before, a demand that remained even after the men had returned home after the war. In 1950, around 8.7 million people in the UK were working in factories, with 700,000 miners, 170,000 navvies, and

80,000 dockers. And on the land there were 750,000 farm laborers and 300,000 horses. Over half the workforce worked in primary or secondary industries.

The importance of manual labor might seem surprising, but one factor had always limited how many jobs machines could perform. Though they were more precise and more powerful than humans, they were also far more inflexible. Machines could not cope with the variability of the world around them; they could not judge distances, assess the force they would need to exert, or make allowances for the unexpected. Above all they could not think for themselves. They had to be constantly tended by human operators, were restricted to the simplest of tasks, and had to be confined to the predictable environments of factories. Even so, machines had always been prone to getting jammed or malfunctioning and could prove lethal to an operator who was unfortunate enough to be drawn into their workings. Hence the need for laborers to work in the open air and down mines and for skilled machine operators in factories and workshops. The engineering triumph of the last seventy years has been to close the gap between machines and the environments in which they are used, massively expanding the range of tasks that they can perform.

The simplest way to achieve this is to standardize the conditions that machines have to deal with, outside factories as well as within. This technique has transformed how goods are transported and has revolutionized world trade. Before the 1950s moving goods was complex, especially the transferring of them between ships, trains, and trucks. Individual items had to be loaded into the hold of a ship and on arrival unloaded, by large numbers of skilled dockers, who operated the cranes and had to work out where each item should be placed and the order in which they needed to move them. Loading and unloading ships could take days and increased the price of shipped goods by around 20 percent. Two Americans, the trucking magnate Malcom McLean and the engineer Keith Tantlinger, removed all this fuss at a stroke in 1955 by inventing the standardized shipping container. Ports could have a single huge overhead or gantry crane that could transfer the containers rapidly onto and off trucks or railway carriages and onto and off container ships. The cranes could simply lift the eight-foot (2.44-meter) wide, eight-foot six-inch (2.59-meter) high, and twenty-foot (6.10-meter) or forty-foot (12.19-meter) long containers, place them side by side on the deck of a ship, and stack them one on top of each other like building bricks or take them off a ship and place each on the bed of a standard truck or railway carriage. Containerization massively reduced the time to load and unload ships, and the numbers of people this needed, and consequently slashed shipping costs by over 90 percent. Huge new container ships have been built that can transport over twenty thousand containers, and sea traffic has moved from small inland harbors to huge new coastal ports, which employ far fewer dockers.

The packing and unpacking of the containers themselves, meanwhile, has shifted to inland warehouses and distribution centers where labor and land costs are lower. Manpower has further been reduced thanks to the standardization of the packages that the containers hold. Goods are mounted on standardized pallets, measuring forty-eight by forty inches (1,219 by 1,016 millimeters), which can be moved in and out of the containers and onward for local delivery using electric forklifts with standardized forks. With the standardization of both containers and pallets, for the last sixty years shipping costs have been falling inexorably. Thus it has become more economic to manufacture goods in developing countries where wages are lower and regulation less strict, and to transport them around the world to developed nations. This has led to the march of globalization and to the industrialization of Southeast Asia, China, India, Mexico, and Brazil. And as a direct consequence, the last sixty years has also seen the concurrent deindustrialization of the West and the decline of many former manufacturing centers.

It proved more difficult to replace laborers who were working in the highly variable environments of forests, farms, quarries, and mines. It is no mean feat to design machines that can do the jobs of workmen wielding scythes, pickaxes, and handsaws. The reason was not because machines were unable to access these areas; as early as the beginning of the twentieth century, engineers had developed tractors that had huge grooved tires and excavators with caterpillar tracks that could move freely even over the softest and roughest of ground. The difficulty was in mounting tools onto those vehicles and transmitting power to them: gears and cables are clumsy and inflexible; the tools are all too easily damaged by shocks if the equipment runs up against stones or other unexpectedly hard objects; and they can pass on those shocks and damage the motor. The solution was to reintroduce a technology that by the end of the nineteenth century had seemed to be obsolete, and which is still underappreciated—hydraulics.

Even as hydraulic power systems in docks and cities were being replaced with electricity networks, engineers were realizing that hydraulic technology was ideal to transmit power within individual vehicles. As ships grew ever larger and faster, they became impossible to steer using the traditional handheld ship's wheel. Marine architects therefore diverted some of the power from the engine to drive pumps and accumulators to supply high-pressure water to move the rudder. They could also use the hydraulic system to power other parts of the ship, such as the capstans used to raise and lower the anchors and pull in their mooring lines. As the guns of warships got larger, hydraulics supplied the power to rotate the turrets and raise and lower the guns. As airplanes became faster and flew at greater altitudes in the 1930s, engineers also started to fit them with hydraulic systems that could move their flight surfaces, raise and lower their retractable undercarriage, and power the bomb doors and gun turrets of military aircraft. Hydraulic systems also found their way into cars; our brake systems act just like Joseph Bramah's early hydraulic presses, transmitting and magnifying the small forces we apply to the brake pedal to produce large forces on the brake pads; many vehicles also have hydraulic power steering. The main difference between these modern systems and the nineteenth-century ones is that the engineers have replaced water as the hydraulic fluid with inert mineral oils, which overcomes the problems of corrosion and freezing.

Hydraulic systems were the key technology that finally turned tractors into safe machines that could perform useful work around the farm. The early tractors had been supplied with a simple tow bar to attach implements such as a plow, so they pulled it along just like a team of heavy horses. However, if the plow hit a stone or dug too deeply into the soil, it would resist being pulled forward much more strongly. This was no difficulty if you were plowing with horses; they would simply stop and wait for the plowman to sort things out. But with a tractor, the engine would keep the wheels rolling, and short tractors, such as Henry Ford's Fordson, could turn over backward, crushing the farmer beneath them.

In 1926 an Irishman, Harry Ferguson, overcame the problem with a new attachment method, the three-point hitch, in which the tools were rigidly attached at three points to a plate, which was in turn joined to the back of the tractor by a hydraulic linkage. This arrangement meant that a wide range of tools could be rigidly attached to the tractor and raised above the ground so they could be driven to the field, before being lowered into the ground for use. Another benefit was that the weight of the tools moved the center of gravity backward, so the large rear wheels of the tractor could apply more driving force. Finally, the arrangement meant that Ferguson could introduce a hydraulic mechanism, draft control, onto the upper attachment point, so that if the plow encountered greater resistance, the hydraulic system would automatically raise the blades. Farmers could also use the hydraulic system to power a wide range of tools that were driven by the hydraulic-power takeoff: tools such as reversible plows, seed drills, muck spreaders, mowers, swathers, and balers. Ferguson's system became popular in the 1940s, when it was incorporated in the famous Ford-Ferguson tractors. This started a precipitous decline in the number of laborers and heavy horses on farms.

Farms finally became fully mechanized with the introduction of that other archetypal farm vehicle, the combine harvester, which is a mower, thresher, and winnower all in one. Huge mechanical combines had been used on the Great Plains of North America from the middle of the nineteenth century, pulled by large teams of horses, but gasoline- and diesel-powered machines started to appear only in the 1930s. From the 1940s onward they really started to become popular, small machines being attached to the rear of tractors. But from the 1950s onward large self-propelled versions were developed, their machinery being driven via hydraulic systems from the powerful main engine. Rather than being a communal activity, involving the whole rural populace, harvesting became a one-man operation.

Hydraulics also enabled machines to finally replace the thousands of navvies and miners who had built the modern world with their pickaxes and shovels: mining coal; quarrying rock; digging foundations; and moving earth. Toward the end of the nineteenth century, engineers had started to develop steam shovels, which helped excavate the land to build the Panama and Manchester Ship Canals, but these machines were heavy and hard to move about; the ones used for the Panama Canal could only move along their own dedicated railway line. The first self-propelled excavators, powered by steam, were introduced as early as 1897 by the Kilgore Machine Company, but the pulleys and wires used to operate them proved dangerous and unwieldy, and it was difficult to produce a machine with multiple joints like a person's arms.

Consequently, not until the early 1950s did earthmoving machinery powered by diesel engines and using hydraulics to power their arms start to be produced commercially. Manufacturers such as Caterpillar

in the USA and JCB in the UK started to produce a wide range of excavators, diggers, and backhoe loaders, in a wide range of sizes. The
main diesel engine of these machines typically powers three hydraulic
motors. Two of them move the arms of the digger with their powerful
hydraulic rams, like muscles, but pushing as well as pulling, while the
third powers the finer-movement control. Today no quarry, building
site, or farm is without these versatile machines, each of which can
replace the labor of many men. Once the diggers have removed material, they can load it into gigantic diesel-engined tip-up trucks that can
take the material to its destination, where their hydraulic rams lift the
front of the open box and let the material slide out of the back. Heavy
machinery in open-cast mines can strip the land at unprecedented
speed. And in underground mines, workers now use hydraulically operated walking-roof systems to support the coal face as the miners drill
through the seam.

But the technology that has had the most obvious impact on the
modern world is electronics. We are all familiar with the transformative effect that digital computers have had, enabling us to process and
communicate unheard-of amounts of information. However, on its
own this would not have materially changed our world to anywhere
near the extent it has. Electronics has played an even more important
role in helping us control our machinery and enabling it to interact in a
more flexible way with the world around it. Increasingly machines no
longer need people to control and operate them.

Computers have a longer history than we usually realize. They
were being used as long ago as the 1940s, when they were still dependent on thermionic valves for their logic machinery. In the Boeing
Superfortress, the most expensive military project of World War II,
electronics engineers fitted an analog computer to direct the fire of its
gun turrets. Though a gunner optically sighted the guns, the computer
increased his accuracy by compensating for the bomber's airspeed,
the movement of the enemy craft, and other factors. Programmable

digital computers, such as the American ENIAC and the British Colossus, were used to calculate the ballistics of shells and to break signal codes respectively. However, the invention in 1947 in Bell Labs of the transistor, and the innovations of the silicon MOSFET transistor and integrated circuits, enabled computers to be miniaturized, so that they were far more powerful while being smaller, using less energy and being much more reliable. Rather than just powering simple communications devices and large mainframe computers, the progressive power of smaller and smaller chips has enabled them to drive a far-wider range of devices and control a wider range of operations. They were not only able to solve mathematical problems but increasingly to interact directly with the material and engineering world. Combining computers with electronic sensors and servomotors meant that they could start to act as the brains of simple machines.

From the 1980s onwards, computers started to be used in manufacturing: controlling electrically and hydraulically operated robots that could undertake standard tasks in production lines that had previously been done by workers: welding, drilling, riveting, sanding, and painting. Today there are some 4 million such robots worldwide, increasing the speed, reliability, and output of the production lines on which they work. And in the last decade they have been joined by computer-controlled 3D printers, which can lay down plastics and metals to produce components with complicated shapes. These machines reduce the need for the many reduction processes—milling, cutting, and drilling—which have taken up so much of the time and energy in manufacturing.

Improvements in sensors and motors have also enabled computers to escape from the factory. Machines and systems that used to be operated manually or used mechanical controls started to incorporate computer controls. The changes happened first in industry and research, but increasingly the process has moved into distribution centers and into our homes. Refrigerators, washing machines, and central-heating systems may look the same as they used to, but most incorporate chips to program their complex operations, while the internal combustion engines in our cars are fitted with electronic diagnostics that continuously work to optimize performance and detect faults. Even car tires are fitted with devices to detect and warn of punctures. The electric motors in our appliances and electric cars are far more sophisticated devices than traditional ones, having replaced the traditional commutators and brushes with computer-controlled switching.

Computer-controlled machines have even moved into the open air. Foresters are nowadays using complex hydraulically powered harvesters to fell and process the wood of conifers in one mesmerizing operation. A huge multipurpose arm first grips the trunk before cutting through its base with its chain saw and laying the tree down. Then another tool strips the trunk of its side branches and finally chops through the trunk to produce optimally sized logs to be loaded by a grapple skidder and moved to a sawmill. Computers have also allowed us to make effective use of the satellites that our rockets have been lifting into orbit since the 1950s. Satellite navigation systems such as GPS are being linked not only to navigation devices but directly to farm machinery, enabling them to cultivate the land and manage and harvest crops with pinpoint accuracy without needing to be driven by a human. GPS-controlled collars can be used to corral animals electronically to move them around pastureland, without the need for fences, sheepdogs, or cowboys. Computers have been used to control our weapons, from cruise missiles to drones, allowing us to kill our fellow humans from the comfort of our own base. And on-board computers are increasingly being used to control airplanes, ships, and trains and are starting to direct road vehicles such as buses, trucks, and even our cars. In time they may replace the hundreds of thousands of professional truck, coach, bus, and taxi drivers.

So though it might look similar, our world today has been revolutionized from that of the 1950s. For a start it is much richer. The development of containerization, hydraulics, and computer control has enabled us to raise productivity and wealth to unprecedented levels. The mean global GDP per person has been raised at a faster rate than at any other time in history, quintupling from $2,400 in 1950 to $12,000 now. Since there are about three times as many people on the planet now, 8 billion, compared with 2.5 billion in 1950, human output has risen fifteenfold, and following the development of much of Asia and South America, the wealth is spread over a far greater area of the planet.

Unsurprisingly, this meteoric rise in output has yet again resulted in an increased demand for power. Global energy use is now six times as high as it was in 1950. Engineers have for the most part met the demand by using increasingly powerful and sophisticated machinery to mine coal and drill for more oil and gas. Not only do we use gigantic hydraulic excavators for mining and quarrying, but drills for oil and gas incorporate hydraulic pumps that bend the drill head sideways to extract far more of the available fuels and enable fracking shales for gas to become economic. And the oil, gas, coal, and nuclear power stations now supply energy to extremely high voltage national and international grids, so that people can have access to electricity throughout the countryside as well as in towns.

But though power is now transmitted more widely, paradoxically humanity itself has become less widely distributed. Farm machinery has emptied rural areas of much of their population, so the world population has become far more urban. Over 80 percent of people live in cities in the UK and the USA and over half globally. And because of mechanization, far fewer people are employed on farms or in factories, and most of those are employed in the developing world. Today, only around 1 percent of the workforce in developed countries works in agriculture and 10–12 percent in manufacturing. The factory where I had my first holiday job in the 1970s assembling electric switches for MG car doors has long since closed, along with most of the other firms on the trading estate. Most goods are now produced in the Far

East by robots. Employment in manual industries has been hardest hit; not only are there fewer farm and factory workers, but navvies, dockers, and heavy horses have moved into the twilight of history. Today we can control our massive machines and power networks with a few flicks of our fingers and are even developing ways in which we can get the machines to control themselves. The machines have taken over from human labor.

Roman *bestiarius* confronting a lioness. Painting from the amphitheater at the Museo Nacional de Arte Romano, Spain. An armed and trained gladiator was more than a match for any animal.

Museo Nacional de Arte Romano/Courtesy of Wikimedia Commons

A western chimpanzee using a rock to crack palm nuts. Note the short distance the chimp has to accelerate the rock and its insecure, hooked grip.

Anup Shah/Alamy

The first blunt force murder? Cain killing Abel. Note the long distance over which Cain can build up energy in the stone, as he rotates his shoulders and swings his upper arm and forearm in a two-stage sling action.

From Gerard Hoet's *Figures from the Bible* (1728)/Courtesy of Wikimedia Commons

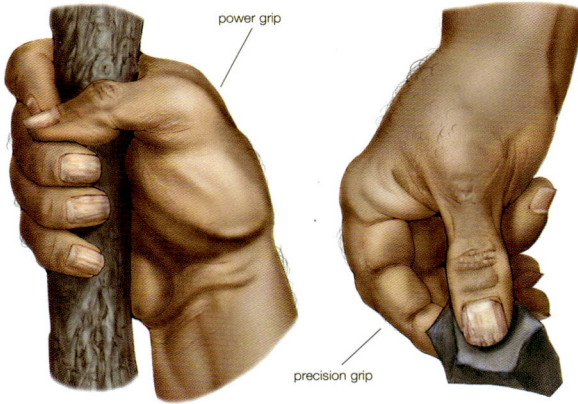

The power and precision grips used by early hominins to grasp large stones or branches and small stone flakes, respectively.

A Lower Paleolithic flint hand ax from Norfolk, England. This large tool fits comfortably in the hand using a power grip.

Stone blades and flakes from the Upper Paleolithic, northern Budapest, Hungary. Most appear to be knives or scrapers but the flake in the center is a burin used to incise grooves or drill holes in antler or wood. All of them have to be held in a precision grip or be hafted onto a wooden handle.

Scenes of ancient Egyptian cereal cultivation from the tomb of Nakht. Note the use of many sling-action hand tools: mattocks to break up the soil; an ax to fell a tree; mallets to break up soil clods; and a whip to encourage the plow oxen, which provide the only other source of power.

Metropolitan Museum of Art/Courtesy of Wikimedia Commons

Two men threshing a sheaf of wheat with flails, one of the most labor-intensive tasks involved in cereal farming. From the Luttrell Psalter, c. 1330.

British Library/Courtesy of Wikimedia Commons

Scene of a gardener using a shadoof, the first of many machines to raise water, tomb of Ipuy at Deir-el-Medina. The only energy used is to lower the container against the counterweight.

Norman de Garis Davies/Courtesy of Wikimedia Commons

A reconstruction of a medieval counterweight trebuchet at the Château des Roure, Ardèche, France. Note the sheer scale; the trebuchet is 21 yards high and has two treadwheels, within which the operators walk to load it.

EMLACH/Courtesy of Wikimedia Commons

The famous line of windmills at Kinderdijk, the Netherlands, built in the eighteenth century to drain the polders. Note the rotating cap of the mills and the twist in the sails that improves aerodynamic efficiency.

Alf van Beem/Courtesy of Wikimedia Commons

An Iron Forge by Joseph Wright of Derby, 1772. Note the water wheel on the left, which is raising a trip hammer with a cam to beat the white-hot iron. Wright was a master of painting artificially lit interiors.

Tate Britain/Courtesy of Wikimedia Commons

The huge Quarry Bank Mill at Styal, near Manchester, England. Note the rural location beside the River Bollin, which provided the 200 horsepower needed to drive the machinery inside.

Francis Franklin/ Courtesy of Wikimedia Commons

The *Rocket* of 1829, built by George and Robert Stephenson. Hot smoke from the firebox is drawn through tubes in the boiler. High-pressure steam enters the inclined cylinders, pushing the piston rod between sliders to drive the crank. The expelled steam rushes up the chimney, drawing with it the smoke.

GRANGER Historical Picture Archive/Alamy

Interior of the Crystal Palace, 1851, a masterpiece of prefabricated glass, iron, and wood. Note the full-sized elm tree around which the structure was built!

Victoria and Albert Museum/Courtesy of Wikimedia Commons

National Gas Engine 1908 Type S Anson at the Anson Engine Museum, Poynton, Cheshire, England. Gas engines were arranged like a steam engine, with a single large cylinder driving a flywheel. Early oil, gasoline, and diesel engines looked much the same.

Clem Rutter/Courtesy of Wikimedia Commons

Tower Bridge, London, the most famous hydraulic machine in the world. Steam engines and a weight-loaded hydraulic accumulator raised and lowered the 1,200-ton bascules in under five minutes.

Mvkulkarni23/Courtesy of Wikimedia Commons

Cyrus McCormick's reaping machine of 1831 heralded the rise of American agriculture and industry, starting the inexorable fall in the rural population culminating a hundred years later with the combine harvester that both reaped and threshed the crop.

World History Archive/Alamy

The electricity hall of the Chicago World's Fair, 1893, demonstrated the power of Westinghouse's AC supply system. For the first time an exhibition could stay open all night.

Ruby Loftus Screwing a Breech-ring by Dame Laura Knight, 1943. Even in factories, the electrically powered machines needed to be operated by people, hence the large number of women working in the background.

A large hydraulically operated dump truck driving on the floor of a copper mine, Zambia, Africa. Note the steps for the driver to ascend to the cab. The largest trucks can shift 450 tons of rock, replacing ten thousand workmen.

Computer-controlled and hydraulically powered robots perform welding operations at the production workshop of Great Wall Motor's Taizhou Smart Factory in Taizhou, Jiangsu, China. Note the total lack of people.

Cynthia Lee/Alamy

Global primary energy consumption by source

Primary energy is based on the substitution method and measured in terawatt-hours.

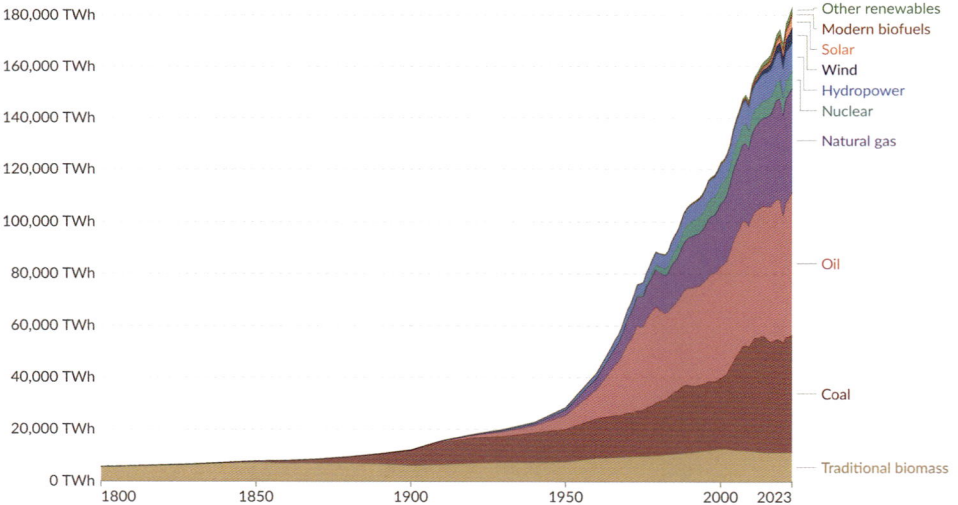

Other renewables
Modern biofuels
Solar
Wind
Hydropower
Nuclear
Natural gas
Oil
Coal
Traditional biomass

Data source: Energy Institute - Statistical Review of World Energy (2024); Smil (2017)   OurWorldinData.org/energy | CC BY
Note: In the absence of more recent data, traditional biomass is assumed constant since 2015.

The rise in global energy consumption. Notice how the figure has risen sixfold since 1950 and is continuing to increase, as is the use of fossil fuels, even with the rise of renewable sources.

OurWorldinData.org. Ritchie et al. (2020). Smil (2017). CC BY

Chapter 19

REINING IN OUR POWER

We have come a long way in the last ten thousand years. Using the physical attributes and the engineering skills that we inherited from our hunter-gatherer ancestors, we have developed increasingly sophisticated ways of manipulating the world about us. We produce more food and worldly goods than at any other time in history and have developed infrastructure that supports an unprecedented number of people in unparalleled ease, health, and plenty. But in doing so we have damaged not only the planet on which we live and rely, and the organisms we share it with, but ourselves.

Unfortunately, in our efforts to feed a population that has risen a thousand times, from 8 million to 8 billion, we have taken over vast areas and left less for the natural world. The Dorobos we met in the prologue are hunter-gatherers and have modest needs; they were content to steal just a single joint of meat from the pride of lions. To meet our demands for our own food, we cultivate 6 million square miles (15.5 million square kilometers), 15 percent of the habitable area of the earth. But we do not even eat all the food we grow but feed a quarter of it to our domestic animals, and we set aside a further 12 million square miles (31 million square kilometers) to grow grass to feed them. And all so that we can produce a glut of our favorite luxury food, meat. Since the efficiency of converting plant biomass into animal biomass is only around 10 percent, the 75 percent of the land we cultivate to produce meat provides less than a quarter of the calories we consume. Altogether we have stripped 18 million square miles (47 million square kilometers), 44 percent of the world's natural vegetation, for our sole benefit. Since 1970 alone, 73 percent of global wildlife has been lost as our population doubled.

And we are a consumer and destroyer of much more than land; we are still ramping up our plunder of the earth's resources and are depleting them at an unsustainable rate. For centuries, many natural resources were able to regrow and replenish themselves as fast as we removed them. We harvested timber and wood using sustainable arboriculture and coppicing. The bounty of the sea quickly restored fish stocks. But in the last sixty years, powerful hydraulic machinery and lightweight chain saws have enabled us to access and destroy virgin forests faster than they can regrow. We fell them to provide timber or to simply clear the land to produce more feed crops for our domestic animals or cheap vegetable oils and sugars for our processed food. Meanwhile giant trawlers scrape off huge areas of the seafloor and scoop up shoals of fish faster than they can reproduce, much of it destined once again to be used to feed animals. Forests and fish stocks are both shrinking fast.

Our increasingly sophisticated machinery has also enabled us to mine and quarry more minerals in the last fifty years than in all of our previous history put together; in the last twenty years alone, the rate of extraction has risen by more than half. And while we plunder the earth, we also damage it faster than it can heal. In the past, nature could cover over its scars, as we used to demonstrate to our students in my department at the University of Manchester. Northwest England was well and truly trashed in the nineteenth century, yet nature there ultimately triumphed over our pollution. Members of our department documented the demise of the melanic form of the peppered moth and recovery of the typical mottled form after the clean air acts of the 1950s reduced the amount of coal people burned and as tree trunks and walls became less sooty. We tracked the return of snails to the city following the reduction in acid rain. We took students to former industrial sites such as brickworks, glassworks, and soda works and showed them the diversity of orchids and salt-marsh species that were colonizing and gradually detoxifying them. We showed them around the derelict land used by Manchester's largest sewer works that my partner's son converted into a local nature reserve. We monitored the recovery of the River Medlock, once an open sewer, and the Manchester Ship Canal, once so full of chemical waste it would periodically burst into flames.

Even large modern dumps of pollution can be returned to nature. Driving along the M62 motorway, I regularly pass a hill that rises from the dead-flat Vale of York. With an area of 1.2 square miles (3 square kilometers) and a height of 225 feet (69 meters), Ash Hill looks as if it has always been there, but it was built only since 1965, made from the pulverized ash that was pumped there for forty years from the nearby coal-fired power stations that produced 15 percent of the UK's electricity. The coal-fired power stations are now closed, and the hill has been planted with grass, hedges, and trees, so now it now looks like a quintessentially English tract of countryside.

In the last fifty years, however, we have exported most of our industry and pollution to the developing world, so there are much larger scars on the landscape that we in the developed world choose to know little about, and which will take far longer to heal: vast open-cast coal mines, oil sand quarries, and garbage dumps. The pollution we produce nowadays can also have not just a local but a global impact. Toward the end of the last century, chlorofluorocarbons (CFCs) from refrigerators built up in the earth's upper atmosphere, enabling cosmic rays diverted into the polar regions by the earth's magnetic field to destroy the ozone that shields us from UV radiation. There had to be an international ban on the chemicals to allow the ozone layer to start to recover.

The root cause of our destructiveness lies in our system for producing our food—cereal farming. Despite thousands of years of effort to improve them, our crops are still relatively unproductive and hard to cultivate. And ever since we started farming cereals, and especially since the advent of industrialization, we have sought to solve these problems using methods that use more energy, so we constantly face pressure to generate more power. This in turn causes more problems that we can only solve by generating yet more power. We are stuck on an accelerating merry-go-round of conspicuous consumption and destruction.

Take cereal farming itself, for instance. To grow more crops while using less effort, we have progressively harnessed the power of draft animals to cultivate the land, waterwheels to grind grain, and diesel-powered tractors and combine harvesters to harvest and thresh the grain. We have been able to minimize our human labor to grow our crops, but we use much more energy to do so. Together the energy our machines use to cultivate the ground and harvest the crops; the energy we use making and applying chemical fertilizers, pesticides, and herbicides; and the energy we use to transport and process our food are greater than the energy we obtain from eating it. Agriculture is now a net consumer, not producer, of energy, using around 30 percent of the total we expend.

The same is true of our manufacturing industries, which over the past few centuries have replaced power from human muscles with that from horses, water, wind, coal, gas, and oil. Today industry uses some 37 percent of all the energy we expend. And we use more and more energy in our daily lives, just to keep ourselves warm, fed, and clothed and to travel around. We also have more and more possessions; the average weight of house moves in the United States is 8,000 pounds (3.6 tons). Even our entertainments and pastimes have become vastly more energy intensive. In medieval times, a "holy day" involved beer, skittles, and dancing at the local church; in the nineteenth century people upped their expectations to a train excursion for a few days at the coast; nowadays people expect long-haul flights to luxury tropical, all-inclusive resorts. In 2022, each citizen of the UK used on average ten times more energy than they needed to maintain their metabolism; every American, over twenty-five times.

We have been so extravagant in our energy use that in the 1970s people used to worry that we would run out of fossil fuels. Since then, however, enhanced methods to drill for oil and frack gas have ensured that we are extracting and burning more of it than ever. However, we face another, far greater challenge: the pollution we have released by burning fossil fuels. We continue to produce huge volumes of carbon dioxide, which we pump out of our homes, cars, planes, ships, factories, and farmland into our celestial cesspit—the sky. The rising concentration of carbon dioxide in the earth's atmosphere is enhancing its natural greenhouse effect, raising global temperatures, and causing destructive changes in weather, driving heat waves, droughts, floods, hurricanes, and forest fires. As the carbon dioxide is absorbed into the oceans, it is also acidifying the seawater, harming sea life that is already affected by the higher temperatures. We face a global environmental catastrophe.

And while our profligate use of land and energy continues to ruin our planet and threatens our existence, it is not even making us any happier. The mechanization of cereal farming and industrialization has taken from us much of what makes us distinctively human and gives us satisfaction and contentment in our lives—the opportunity to marry our physical and mental skills, to use our hands and bodies to feed, clothe, and house ourselves. Cereal farming enslaved us to backbreaking labor on the land. Gradually, our engineering ingenuity reduced these demands. The first phase of industrialization doomed us to be the slaves of machines or forced us to mimic them, carrying out repetitive but unskilled tasks; the last fifty years have left many people without any work at all as we have been supplanted by computer-controlled machinery. The economist Adam Smith recognized the problem back in the eighteenth century; he had little to suggest other than mass education, but did not explain what well-educated unemployed people could do.

Knowing little about the engineering that underpins our society, members of today's chatterati have little idea either, other than exalting the role of the arts—usually the one for which they themselves are subsidized—and suggesting that more people should take part in them. But not everyone has the ability or inclination to act, make up stories, or create images. Some people can excel at games, and the last few decades have seen a meteoric rise in the importance we place on the actions of professional sportspeople. Many people, however, get fulfillment by doing more useful things. I have always been struck by the contentment of those who make their living through their practical skills: technicians, builders, plumbers, electricians, chefs, gardeners, and tree surgeons, for instance, who lead useful, meaningful lives and meet real needs. People in jobs that have replaced work in factories or farms—in offices, call centers, and distribution centers, for example—report far less job satisfaction.

For the last forty years there has been much talk about how we can avert the climate disaster by "transitioning" from fossil fuels to renewables. We have started to update old technologies, using wind and water power to supply electricity, and we have been developing new ones, such as solar panels that generate electricity, and lithium-ion batteries that can store it so we can power our vehicles. Many people expect, or hope, that by harnessing such engineering brilliance, technology can save us. However, my experience causes me to doubt this. When I took up my first permanent academic post in Manchester in 1990, some of my research colleagues were already monitoring the effect of climate change in the Arctic, which was warming fast, just as climate models were predicting. And when in 2000 I started to investigate the potential of tree planting to cool down cities and reduce flooding and so adapt them to climate change, the models we used predicted changes in UK weather in the 2020s that have proved all too accurate. We knew all about the problem decades ago. Yet there has been little effective action to reduce CO_2 emissions. Despite a rise in the use of renewables, energy production, fossil fuel use, and CO_2 emissions continue to increase as poorer nations develop and the world population rises.

It seems unlikely that things will change anytime soon. Multinational oil companies will continue to cast doubt on climate research and scaremonger about the risks of transition; and people will continue to be unwilling to vote for political parties that allow solar farms and wind turbines to be built near their homes and that promote policies that might reduce their living standards. It suits everyone to carry on pretending that we are taking action while essentially doing nothing. After all, much of the destruction is taking place in remote, sparsely inhabited, and poorer parts of the world, such as the Arctic and small coral islands, so it can conveniently be ignored. Only massive disasters affecting the rich and powerful, such as the Great Stink that assailed Britain's parliamentarians in the 1850s, and which finally prompted them to clean up the River Thames, are likely to prompt real action.

But even if we do succeed in replacing fossil fuels by renewables to produce electricity, it will be far harder to eliminate carbon dioxide emissions from the manufacture of plastics, steel, and concrete. And if we do finally manage that more difficult task and become "carbon neutral," it will still not overcome all of our problems. For a start, we will continue to wreak devastation on the planet. Currently we are using up the earth's resources at 1.7 times the rate at which they can be restored, and people in the developed world are using more than that. In the UK the figure is 2.6 times, while in the US it is 5.1 times. We will continue to practice intensive cereal farming and deforestation, and we will add other ways of damaging the planet. There will be a greater demand, for instance, for the minerals used in electronics and in the new carbon-free technologies: minerals such as lithium, cobalt, and the rare earths. We will continue to destroy the forests and empty the seas. We will continue to destroy our environment and will only take its plight seriously, as one of my Chinese students succinctly put it, "when we have ruined it."

To survive and thrive on this planet, we need to reverse our direction of travel. Rather than trying to solve our problems using methods that use more energy, as we have been doing for the last ten thousand years, methods that inevitably create more problems, we need to remove our problems by reducing the energy we use—by reining in our power. And there is an alternative way in which we can support ourselves, whether we choose to take it up before we destroy our environment, or whether we wait until after a global catastrophe forces it upon us. It is one that could make people happier, reduce our impact on our planet, and lead to a world that is truly sustainable. We need to return to an earlier stage in our history when we were not so destructive.

Obviously, we could not go back all the way to our original hunter-gatherer lifestyle, since we could not harvest enough food to support the 8 billion people who live on the world today. Most mature ecosystems have a high standing biomass but a low productivity; trees have reached their maximum height, and most plants defend their tissues with inedible fibrous tissues or expensive defensive chemicals. However, as we saw in chapter 8, there is a second alternative—horticulture. The yields of root crops, vegetables, and fruit trees are many times as great as those of cereals, so cultivating them in small plots can support high population densities. The Central American empires were fed entirely by horticulturalists tending their gardens using only manual labor, with no help from draft animals or machinery. Similar systems can be and have been successful elsewhere. The market-gardening system developed in the fifteenth-century Netherlands and the mixed horticulture and agriculture systems developed in China and seventeenth-century England enabled them to feed rapidly growing populations.

And since the seventeenth century, the New World crop plants that have been imported into Europe have helped support yet higher populations and vastly improved the diets of Europeans. The cultivation of tomatoes, peppers, zucchini, sunflowers, and sweet potatoes, together with eggplants and citrus fruits imported from Asia, have formed the basis of the healthy Mediterranean diet.

The superiority of New World crops over cereals and their impact on the Old World is best embodied in the humble potato, which hopped over to Europe via the Canary Islands in the sixteenth century, and which Europeans finally started to cultivate in earnest in the eighteenth century. Grown in peasants' gardens, potatoes helped banish the threat of famine and enabled the poor to survive the incursions of invading armies, who had hitherto found it all too easy to steal their stores of grain. With a yield five times that of wheat and a high, balanced protein content, potatoes also fed the growing population of manufacturing towns in the nineteenth century and enabled mill owners to keep their wages low, thus increasing their profits and prolonging industrial expansion.

Nothing better exemplifies the superiority of the potato than the case of Ireland. In the early nineteenth century, Irish peasants, banished from the best farmland by their English colonists, had to scrape a living from tiny plots of unproductive bog. Nevertheless, by cultivating a few rows of potatoes in lazybeds and by keeping the odd cow, 8.5 million people sustained themselves in good health. Many emigrated to Great Britain, where they used their unparalleled physiques, working as navvies to build an industrial nation. The tragedy of the Irish was that they were growing a monoculture of Lumper potatoes, which in the 1840s were devastated by potato blight. Without sufficient aid from the British government, a million people starved, a million others were forced to emigrate, and the population halved in just a few years. It has never regained that figure, and the present population of Ireland is only around 5.3 million.

Today, with greater knowledge of agronomy and plant pathology, we could easily grow enough food to feed ourselves in a sustainable way by cultivating a broad diversity of plants and keeping hens, pigs, and other small animals to vary our diet, recycle biomass, and help maintain soil fertility. Allotment holders using conventional gardening techniques, organic growers, permaculture enthusiasts, and forest gardeners can all produce many times more food per unit area on their plots than a cereal farmer. Assuming an average productivity that is four times as great as that of modern cereal farmers, we would only need around 1.2 million square miles (3 million square kilometers) of land worldwide to grow the vegetables we require, just 20 percent of what is currently under the plow, and 7 percent of the total farmland. Even if we also set aside some land to grow crops for animals so we could maintain our meat consumption, we would still need less than 12 percent of current farmland—leaving a far-greater area for forests and grasslands and the plants and animals that they support. Natural ecosystems would regenerate, just as they did in the Yucatán Peninsula following the demise of the Mayans, or in the Amazon after the Columbian invasions decimated the local population.

Using composting and manuring to maintain soil fertility rather than chemical fertilizers, and using hand tools such as spades and hoes rather than plows and combine harvesters, horticulture would use far less energy than intensive agriculture. It would also be far better for our physical and mental health. After all, the people who have the most choice about what they do with their time—the retired—take to horticulture in huge numbers, cultivating their gardens and allotments and working as volunteers to provide much of the labor force for public open spaces, community gardens, and the grounds of historic houses. As someone who volunteers myself at our local botanic garden, I can also report that gardening brings great benefits. Undertaking natural open-air exercise keeps us physically fit and healthy, while being absorbed in manual tasks reduces stress and leaves us contented. With our plants to think about, and our physical fatigue, we have less need to distract ourselves with extravagant holidays or expensive entertainments. The energy input into food cultivation and other industries would fall dramatically, while the carbon stored in the regenerating ecosystems would rise. Carbon dioxide levels in the atmosphere would start to level off and eventually fall.

Persuading people to convert to horticulture and splitting up agricultural land into smaller garden plots would be hard policies to sell to a population who have been brought up to a life of plenty, ease, and comfort. Many, if not most people, would be loath to give up comfortable desk jobs for hard physical labor in all weathers. However, when there is no alternative, people can readily respond to the challenge. The breakdown of the economies of Eastern Europe following the fall of communism, and the calamitous fall in the value of state pensions, led people in many of the countries to revert to the traditions of their forefathers: to cultivate large gardens, produce their own fruit and vegetables, make pickles and jams, and even brew and distill their own alcoholic drinks. Like the nineteenth-century Irish, they remained healthy and lived surprisingly well. And in World War II, as the United Kingdom faced starvation, the government freed up more land so that individuals could grow their own fruit and vegetables. As part of the Dig for Victory campaign, they created 1.3 million new allotments, covering 130 square miles (337 square kilometers) of land. Even the moat of the Tower of London was converted to vegetable plots. Despite making up less than 0.3 percent of the total land area, the allotments enabled people to grow an extra 1.4 million tons of food a year. This contributed 15 percent of the reduction in the amount of food the country needed to import: down from 22 million tons at the start of the war to 12 million tons at the end. And this was at a time when millions were away at war and the rest were working flat out for the war effort.

The other main barriers to transforming our way of life from cereal farming to horticulture would no doubt be from entrenched economic interests. Farmers would be unwilling to give up their land, even though, because of the low yields of cereal crops, they make poor use of what they own. An even more powerful vested interest would be the agrochemical industry, which makes so much money selling fertilizers, pesticides, and herbicides that enable farmers to continue the unsustainable cultivation of monocultures of cereals and other crops, and the intensive husbandry of cattle, sheep, and pigs. The rest of industry would be unhappy, too, as it would result in a massive reduction in secondary and tertiary industries. With people spending more time and energy growing their own food, they would have less time and money to buy goods and services; the economy would shrink and there would be the anathema of economists: degrowth.

But we would still enjoy the positive legacy of our ten thousand years of cereal farming and hundreds of years of industrialization: our scientific knowledge, engineering know-how, and medical expertise. So not all of industry would be lost, and we would not return to a new "dark age." Indeed, the new way of life would stimulate industry to develop a new range of sustainable machinery to lighten our workload while using energy more economically: a series of electric pumps, strimmers, harvesters, and rotavators powered by solar cells or other renewable sources would help reduce the labor in gardening. No doubt people would find a way of selling other products and services to help them enjoy their lives even better in the new circumstances. We could make use of the regenerating forests to supply our need for timber and coppice the forests for firewood. We could build an economy that is happier, healthier, greener, and more egalitarian than the one in which we are currently stuck. Looking after a small area of land would help free us from the three great evils the eponymous hero of Voltaire's *Candide* and his pals recognized: boredom, vice, and poverty. After centuries of toil, struggle, and war during which we have searched in vain for peace, happiness, and contentment, what we need to do is, like them, cultivate our garden.

Acknowledgments

This book came about as a result of ideas I developed about how we move and wield our tools as I was writing my previous tome, *The Science of Spin*. I would like to thank Adam van Casteren for helpful discussions on the topic and Dr. Lydia Luncz for sharing details of her research on the use of hammers by chimpanzees. Many thanks, too, for my partner Yvonne Golding for many things, not least for indulging me with birthday visits to the Kelham Island Museum, Sheffield; Leeds Industrial Museum, and Bamburgh Castle. Also to Frauke Alves for taking me to the excellent Upper Harz Mining Museum in Clausthal-Zellerfeld, Germany.

I thank my editors, Sam Carter at Oneworld and Colin Harrison at Scribner, for their excellent feedback, particularly in improving on the first draft of the book, and Emily Polson and Hannah Haseloff for taking me through the production process so skillfully. I also thank my agent, Peter Tallack of the Curious Minds Agency, for continuing to believe in my efforts and for his excellent advice.

Most of all I'd like to thank my friend Peter Lucas, whom I first met in the guest house of the Danum Valley Study Centre, Sabah, Malaysia, more than thirty years ago. We have had many useful discussions over the decades about biomechanics, science, and academia. I owe him a particular debt for taking my ideas about primate and human evolution seriously, so that I ever had the gall to start writing books on these topics.

Notes

Most of the information set out in this book, particularly historical facts and statistics, is well-known and documented and is freely available in school textbooks and online, through websites such as Wikipedia. The authors of these works have assembled the facts to produce what has become our accepted wisdom, the stories that we are taught in schools and universities. The accepted wisdom is not static, however. Scientists, archeologists, anthropologists, and historians are continually making new discoveries, especially about the early history of our species, and this gradually shifts our understanding of the past. Many of the notes below, therefore, refer to recent research papers that are starting to alter our understanding of our evolution and prehistory. Notes for works found in the References section are identified by author and year of publication; for full citation see the References section. But there are other reasons why our accepted wisdom may need to be shifted. Historians may have misinterpreted information on technical subjects with which they are unfamiliar or may have failed to consider these subjects altogether. Topics such as science and engineering are usually given only lip service in world histories, and are instead restricted within the silos of "history and philosophy of science" and of "industrial archeology." Consequently, I have found far fewer works to which I could refer in the second half of the book. I have had to use freely available technical information and to rely on my own interpretation of the facts to show how technology has steered the progress of history and the development of civilization.

PROLOGUE: THE POWERFUL PRIMATE

ix *three members of the Dorobo people of Kenya:* https://www.you
tube.com/watch?v=Wo2dxfx-eq4.

x *Recent research by Liana Zanette:* Zanette et al. (2023).

CHAPTER 1: THE DAWN OF POWER

3 *Hedwige and Christophe Boesch of the University of Zurich:*
Boesch and Boesch (1982).

3 *Jane Goodall discovered that the chimpanzees:* For a good description of these discoveries see Goodall (1971).

5 *chew it into a digestible pulp:* See Lucas (2004) for an explanation
of how teeth break down food.

5 *Apes such as chimpanzees can even test whether a fruit is ripe:*
Dominy (2004).

6 *their larger brain also enables them to judge:* See Ennos (2020) for
a broader discussion.

7 *My former PhD student Adam van Casteren showed:* Van Casteren
et al. (2013).

7 *In the swamp forests of Suaq, Sumatra:* van Schaik (2004).

8 *Adam van Casteren was part of a team who showed:* Pascual-Garrido
et al. (2025).

8 *The chimpanzees of Gabon, for instance, carry around:* Boesch et
al. (2009).

9 *The savanna chimps in Tanzania, East Africa, make digging sticks:*
Hernandez-Aguilar et al. (2007).

9 *Females construct spears:* Pruetz and Bertolani (2007).

11 *But nutshells are brittle:* See Lucas et al. (1991) for a description of
the mechanical design of nuts

12 *Chimpanzees who have grown up in a group:* Koops et al. (2022).

12 *In California, the Cahuilla tribe developed a whole "balanoculture":* Logan (2005).

CHAPTER 2: PUTTING OUR BACKS INTO IT

14 *The arm muscles of apes make up a far larger proportion:* Thorpe
et al. (1999).

14 *And their muscles are also composed of a greater proportion:* O'Neill et al. (2017).

14 *Muscle is an extraordinary tissue:* For more information about the capabilities of muscle see Alexander (2003).

16 *Consequently their triceps muscles are 25 percent smaller:* Thorpe et al. (1999).

16 *the nuts the Taï chimps eat:* Boesch and Boesch (1982). Unfortunately, primatologists' understanding of fracture mechanics is sketchy, so in many of their papers they confuse force and energy.

16 *This controlled-dropping technique:* Visalberghi and Fragaszy (2013) and Gumert and Malaivijitnond (2013).

17 *However, there are other ways to swing our limbs:* I introduce these ideas at greater length in Ennos (2023).

20 *Experiments (presumably not on living subjects):* Gläser et al. (2011).

20 *they have been unable to propel stones:* Roach et al. (2013) and Roach and Lieberman (2014). These papers are mostly about javelin throwing in humans, but despite the first being in the prestigious journal *Nature*, they miss the role of sling action in throwing. They end up assuming that javelin throwers must be storing energy in their shoulders. In fact there are no structures in the shoulder to store energy, unlike the elastic tendons of our Achilles' in the heel or of our hamstrings.

20 *Boxers deliver two main types of punches:* For the details of fist speed and time to land a punch see Stanley et al. (2018).

22 *by 5 million years ago creatures such as:* For more details about the evolution of bipedality in hominins see Desilva (2021).

23 *A recent simulation study led by Karl Bates:* Bates et al. (2025).

24 *Species such as* Australopithecus africanus *would have:* For a nice discussion of the transition of hominins from a vegetarian diet to one including marrow and brains, see Thompson et al. (2019).

CHAPTER 3: GETTING A GRIP

28 *Their palms are slender, their fingers are long:* Large numbers of papers have been written about the evolution of the human hand and its difference from that of apes. Perhaps the clearest and most concise summary is that of Young (2003).

29 *Together, these adaptations greatly improve the efficiency:* Young (2003) has some lovely diagrams of people holding all sorts of objects, from baseballs to golf clubs.

30 *we can make far better use of long:* For a longer explanation of how we use clubs to hit things, see Ennos (2023).

CHAPTER 4: CUTTING IT

36 *However, though they are flexible, they are also strong:* For longer explanations of the mechanics of soft biological tissues with collagen reinforcement, see Ennos (2012).

37 *A colleague of mine at the University of Manchester, Phil Manning:* Manning et al. (2006).

39 *This slicing method cuts the food up:* For more details about the mechanics of slicing food, see the excellent article by Reyssat et al. (2012).

41 *In 2010, a team of anthropologists:* McPherron et al. (2010).

41 *at Lomekwi, Kenya, found evidence for an actual toolmaking industry:* Harmand et al. (2015).

42 *the stone-knapping technology that was to dominate:* Plummer et al. (2023).

43 *Recent X-ray investigations of the fingers:* Skinner et al. (2015).

44 *Joseph Henrich has shown, this would lead hominins:* See Henrich (2016) for a longer treatise on the role of social learning in the rise of humanity.

44 *young orangutans require seven years:* Permana et al. (2024).

45 *At birth, only 1.6 percent of the neurons:* Henrich (2016).

CHAPTER 5: THE TRANSFORMATIONAL TOOL

48 *Some anthropologists have suggested that* Homo erectus: Bramble and Lieberman (2004). Note that the endurance-running hypothesis assumes that *Homo erectus* had invented water containers to carry with them. If they had not, they would lose so much water by sweating during their hunts that they would die of dehydration.

49 *The root cause of these problems is:* For more detail about the mechanical design of wood as a material, see Ennos (2012 or 2020).

49 *Modern chimpanzees make their spears:* Pruetz and Bertolani (2007) and Hernandez-Aguilar et al. (2007).

51 *Wood residues found on the blades of hand:* See Keeley and Toth (1981) and Dominguez-Rodrigo et al. (2001) for references to early woodworking.

52 *The maker has to gradually free it:* A nice simple account of the design and manufacture of stone tools can be found in Bordaz (1970), but lots of good information is also online.

53 *which could raise the effective temperature:* Samson et al. (2017).

53 *Tantalizing evidence of just such a hut:* Leakey et al. (1971).

54 *However, anthropologists agree that definitive evidence:* Wrangham (2009) and Gowlett (2016).

55 Homo erectus *also likely used the heat:* The cooking hypothesis is expounded at length by Wrangham (2009).

56 *Later* Homo erectus *devised a more sophisticated:* Bordaz (1970).

CHAPTER 6: MIXING AND MATCHING

57 *modern humans limit the power they apply:* Coe et al. (2022).

58 *Around five hundred thousand years ago humans developed:* For more details of the different techniques of knapping, the Levallois and Mousterian, see Bordaz (1970).

58 *the knapper would first choose a large stone nodule:* A nice animation of the technique can be seen on Wikipedia at https://en.wikipedia .org/wiki/Levallois_technique.

58 *The hominin would have had to grasp a flake:* Key et al. (2021) discuss the disadvantages of holding flakes directly and the benefits of hafting them.

59 *You first have to make:* See for instance Tomasso and Rots (2018).

60 *To attach blades securely, people had to create:* Tydgadt and Rots (2022) present a nice study on the effectiveness of early glues for hafting tools.

60 *that striking a blunt flint tool against a lump:* Stapert and Johansen (1999) present a nice discussion of how flints were used and how the flints used to light fires can be identified.

61 *the archeologists who discovered the Schöningen spears:* For the discovery of the spears see Thieme (1997), and for subsequent analysis see a series of papers in *Journal of Human Evolution* 89.

61 *javelin throwers can accurately hit targets:* Milks et al. (2019).

CHAPTER 7: TYING IT ALL TOGETHER

63 *In her delightful childhood memoir,* Period Piece: For a modern edition, see Raverat (2018).

65 *glued joints are all too prone to fail at low loads:* For a lovely discussion of the mechanics of glued joints, see Gordon (1978).

65 *a range of engineering solutions to produce cutting tools:* For a longer discussion see Ennos and Oliveira (2017).

66 *The benefits of hafting an ax or adze are huge:* See the experiments of Coe et al. (2022).

67 *Laure Metz and Ludovic Slimak of the French National:* Metz et al. (2023).

67 *It takes around two hundred separate steps:* Lombard and Haidle (2012).

67 *In Europe there was a long tradition:* Westcott (1999).

69 *the climactic scene in Khaled Hosseini's bestselling novel:* Hosseini (2003).

70 *By far the best suggestion has come from enthusiasts:* Westcott (1999), 195–99.

71 *Practiced experimental archeologists can deliver:* See the excellent experimental tests carried out by Coppe et al. (2019).

71 *an archer can still store around:* Coppe et al. (2019).

72 *One solution hunter-gatherers came up with:* For more on the development of fire drills see Hodges (1970).

CHAPTER 8: GROWING OUR POWER

80 *recent studies, summarized by Dave Goulson:* Goulson (2019) and Nicholls et al. (2020).

81 *The men of the Hadza tribe, for example, walk:* Pontzer (2021).

82 *Research led by Herman Pontzer of Duke University:* For all the details of their study on the energetics of different subsistence strategies, see Kraft et al. (2021).

83 *One of the largest was found:* Rostain et al. (2024).

91 *it took two hundred man-hours for members:* See https://theprehistoricworkshop.co.uk/emmer-wheat/ for the group's efforts to grow emmer wheat.

92 *So the conventional story that historians tell:* Diamond (1997) is particularly keen to extol the virtues of cereals over fruit trees and root crops, without considering their low yield or how hard they are to cultivate and process.

92 *neither was early cereal farming a disastrous poverty trap:* As has been suggested by Harari (2015), Scott (2017), and Suzman (2022), among others.

CHAPTER 9: REDUCING THE WORKLOAD

95 *a recent study by Irish and German experimental archeologists:* O'Neill et al. (2022).

96 *Excavations at the early Neolithic village of Catalhöyük:* Eddisford et al. (2009).

97 *The innovation that enabled potters:* For a good discussion of the evolution of early ceramic kilns, see Streily (2001).

99 *Grinding alone takes ten to twenty man-hours:* Madsen (1984).

99 *a group of twenty people using ground-stone axes:* Jørgensen (1985).

99 *Experiments that I carried out with:* Ennos and Oliveira (2017).

100 *they helped shipwrights build bigger boats:* For more details see Ennos (2020).

101 *The economist Samuel Bowles:* For more about the rise of inequality among cereal farmers, see Bowles and Fochesato (2024).

CHAPTER 10: CARVING OUT EMPIRES

103 *a group of the feared FARC guerrillas:* https://www.amazonteam.org/the-isolated-tribes/.

105 *The exceptional properties of metals result:* For a nice explanation see Gordon (1968).

108 *The Iceman, Ötzi, for instance:* Fleckinger (2018).

109 *A study by James Mathieu:* Mathieu and Meyer (1997).

109 *Two of the new tools enabled carpenters:* For more on the design and use of traditional woodworking tools, see Beeler (1996).

114 *Bronze Age wheelwrights used a simple but effective:* See Bulliet (2016) and Ennos (2023) for more details about the evolution of wheels.

116 *Given the huge skill and precision required:* For more on early ships see McGrail (1996).

116 *The oldest surviving craft is the funeral vessel:* For more details see Jenkins (1980).

CHAPTER 11: GEARING UP

121 *the place to visit is the Barbegal mill complex:* For more details about the development of machinery in the ancient world, see Landels (1978) and Hodges (1970).

CHAPTER 12: FORGING MILITARY POWER

133 *the two earliest history books:* Modern translations include Herodotus (2003) and Thucidides (1974).

134 *Jane Goodall was shocked to observe a conflict:* See Goodall (2010) for her account of the struggle.

135 *Later studies, however, have replicated Goodall's finding:* See, for instance, Wilson et al. (2014).

136 *Dr. Marta Mirazón Lahr from the University of Cambridge:* Mirazón Lahr et al. (2016).

137 *Analysis of skeletons in a 13,400-year-old cemetery:* Crevecoeur et al. (2021).

137 *The first known incidence of large-scale warfare in Neolithic Europe:* Fernández-Crespo et al. (2023).

142 *But it was the Romans who took military engineering:* For more on Roman siege weapons see Hodges (1970) and Landels (1978).

144 *The first was the crossbow, invented in the fifth century BCE:* Culp (2017).

145 *In the fifth century BCE the Chinese also:* For more on the history of trebuchets see Chevedden et al. (1995).

147 *the technological game changer in warfare:* The most readable account of the development and use of gunpowder is that of Ponting (2005).

CHAPTER 13: RAISING THE POWER SUPPLY

154 *The wealthiest country in Eurasia was China:* For a good introduction to Chinese agriculture see the works of Francesca Bray, for instance, Bray (2004).

157 *Paul Warde of the University of East Anglia estimated:* See Warde (2007) and the discussion of his work in Wrigley (2010).

157 *The Netherlands led the way:* For more on the development of the Netherlands and its influence on England, see Scott (2019).

158 *Jan de Zeeuw of the Agricultural University of Wageningen estimated:* de Zeeuw (1978).

161 *a newly emerging class of yeomen farmers imported:* See Wilson (1965) for a longer discussion of the factors leading to England's emergence as an industrial country.

163 *The first solution was probably the cam:* For a fuller discussion of medieval engineering, not only camshafts and cranks but also water mills and windmills, see Farrell (2020).

164 *by 1550 mills were producing some 550:* Warde (2007).

164 *In its heyday the Wealden iron industry produced:* Wrigley (2010).

165 *providing the heating fuel that allowed its population to quadruple:* For more details of population growth and coal use in England over the early industrial period, see Wrigley (2010).

167 *Coal production was ramped up from:* Wrigley (2010).

169 *Not until Joseph Wright of Derby:* The best collection of Wright's work is housed in the Derby Museum and Art Gallery. Among his portraits are ones of the industrialists Richard Arkwright, Samuel Crompton, and Jedediah Strutt; Erasmus Darwin, the doctor, inventor, and Charles Darwin's grandfather; and Benjamin Franklin, the scientist, diplomat, and founding father of the United States of America.

CHAPTER 14: POWER FOR PRECISION

171 *On March 4 he showed the Scottish diarist and biographer:* The visit and Boulton's speech were recorded in Boswell (1791).

172 *Writing in the same year as Boulton's famous speech:* Smith (1776).

181 *In 1759, Smeaton showed that overshot wheels:* Smeaton (1759). For a broader discussion of Smeaton's scientific and engineering work, see Skempton (1981).

182 *engineers had to come up with a more intuitive unit:* For a nice discussion of the debate see Stevenson and Wassersug (1993).

CHAPTER 15: POWERING AN INDUSTRIAL NATION

190 *steam-powered machine shops that produced:* For more detail about the development of machine shops and precision engineering, see Winchester (2018).

191 *Bramah's flash of genius was to use Pascal's principle:* For more on the hydraulic press and its development, see McNeil (1972).

199 *even greater sensations:* For more on these American manufacturers see Winchester (2018) and Rasenberger (2020).

CHAPTER 16: TRANSMITTING POWER

206 *In 1845, a Newcastle solicitor, William Armstrong:* For more on the first golden age of hydraulic power, see McNeil (1972) and Jarvis (1985).

212 *By the 1850s French and American engineers had perfected:* For more on the development and use of turbines, see Smith (1975 and 1980). It's notable that authors seem to have totally neglected the important topic of water power for most of the last fifty years even while water turbines produce more than 15 percent of our electricity!

215 *Worldwide energy demand had doubled from 1850:* All figures for worldwide energy consumption are taken from the Our World in Data website, https://ourworldindata.org/energy-production-con sumption.

CHAPTER 17: PORTABLE POWER

226 *By 1950, global energy use had doubled:* Data taken from https:// ourworldindata.org/energy-production-consumption.

CHAPTER 18: THE HEGEMONY OF THE MACHINES

229 *Two Americans, the trucking magnate Malcom McLean:* Levinson (2016).

231 *The solution was to reintroduce:* Strangely, though there are hundreds of books and thousands of articles about technical aspects of modern hydraulic machinery, I have been unable to find any work that considers its huge economic and social impact!

237 *The mean global GDP per person:* Economic data are taken from the Maddison Project at https://www.rug.nl/ggdc/historicaldevelop ment/maddison/releases/maddison-project-database-2023.

237 *Global energy use is now six times as high:* Data taken from https:// ourworldindata.org/energy-production-consumption.

CHAPTER 19: REINING IN OUR POWER

239 *To meet our demands for our own food:* Data taken from https://ourworldindata.org/global-land-for-agriculture.

240 *Since 1970 alone, 73 percent of global wildlife has been lost:* IPBES (2019).

243 *In 2022 each citizen of the UK used on average ten times:* Data taken from https://ourworldindata.org/energy-production-consumption.

244 *Many people expect, or hope, that by harnessing:* Fressoz (2024) describes and demolishes this fallacy.

245 *we are using up the earth's resources at 1.7 times the rate at which they can be restored:* Data from https://overshoot.footprintnetwork.org/how-many-earths-or-countries-do-we-need/.

247 *The superiority of New World crops:* For the full story of the history of the potato and its influence on humankind, see Reader (2009).

IMAGE CREDITS

Page 30: Gordon Dylan Johnson/ Courtesy of Openclipart
Page 124: Image Source Limited/Alamy
Page 143: Jim Killock/ Courtesy of Wikimedia Commons
Page 191: Alamy
Page 194: James Nasmyth/Courtesy of Wikimedia Commons

References

Alexander, R. McN. 2003. *Principles of Animal Locomotion*. Princeton, NJ: Princeton University Press.

Bates, K. T., S. McCormack, E. Donald, et al. 2025. "Running Performance in *Australopithecus afarensis*." *Current Biology* 35:224–30.

Beeler, A. W. 1996. *Old Ways of Working Wood*. Edison, NJ: Castle Books, 1996.

Boesch, C., and H. Boesch. 1982. "Optimisation of Nut-Cracking with Natural Hammers by Wild Chimpanzees." *Behaviour* 83:265–86.

Boesch, C., J. Head, and M. M. Robbins. 2009. "Complex Tool Sets for Honey Extraction Among Chimpanzees in Loango National Park, Gabon." *Journal of Human Evolution* 56:560–69.

Bordaz, J. 1970. *Tools of the Old and New Stone Age*. Newton Abbot, UK: David and Charles.

Boswell, J. 1791. Life of Samuel Johnson.

Bowles, S., and M. Fochesato. 2024. "The Origins of Enduring Economic Inequality." *Journal of Economic Literature* 62:1475–1537.

Bramble, D. M., and D. E. Lieberman. 2004. "Endurance Running and the Evolution of *Homo*." *Nature* 432:345–52.

Bray, F. 2004. "Rice, Technology, and History: The Case of China." *Education About Asia* 9:14–20.

Bulliet, R. W. 2016. *The Wheel*. New York: Columbia University Press.

Chevedden, P. E., L. Eigenbrod, V. Foley, and W. Soedel. 1995. "The Trebuchet." *Scientific American* 273:66–71.

Coe, D., L. Barham, J. Gardiner, and R. Crompton. 2022. "A Biomechanical Investigation of the Efficiency Hypothesis of Hafted Tool Technology." *Journal of the Royal Society Interface* 19:2021.

Conway, E. 2023. *Material World*. London: W. H. Allen.

Coppe, J., C. Lepers, V. Clarenne, E. Delaunois, M. Pirlot, and V. Rots. 2019. "Ballistic Study Tackles Kinetic Energy Values of Palaeolithic Weaponry," *Archaeometry* 61:933–56.

Crevecoeur, I., M.-H. Dias-Meirinho, A. Zazzo, D. Antoine, and F. Bon. 2021. "New Insights on Interpersonal Violence in the Late Pleistocene Based on the Nile Valley Cemetery of Jebel Sahaba." *Scientific Reports* 11:9991.

Culp, J. 2017. *Ancient Chinese Technology*. New York: Rosen Publishing.

de Zeeuw, J. W. 1978. "Peat and the Dutch Golden Age. The Historical Meaning of Energy-Attainability." *A.A.G. Bijdragen* 21:3–31.

DeSilva, J. 2021. *First Steps: How Walking Upright Made Us Human*. London: William Collins.

Diamond, J. 1997. *Guns, Germs, and Steel*. New York: Random House.

Dominguez-Rodrigo, M., J. Serrallonga, J. Juan-Tresserras, L. Alcala, and L. Luque. 2001. "Woodworking Activities by Early Humans: A Plant Residue Analysis on Acheulian Stone Tools from Peninj (Tanzania)." *Journal of Human Evolution* 40:289–99.

Dominy, N. J. 2004. "Fruits, Fingers, and Fermentation: The Sensory Cues Available to Foraging Primates." *Integrative and Comparative Biology* 44:295–303.

Eddisford, D., R. Regan, and J. S. Taylor. 2009. "The Experimental Firing of a Neolithic Oven." In *Çatalhöyük 2009 Archive Report: Çatalhöyük Research Project*, edited by S. Farid.

Ennos, A. R. 2012. *Solid Biomechanics*. Princeton, NJ: Princeton University Press.

Ennos, A. R., and J. A. V. Oliveira. 2017. "The Mechanics of Splitting Wood and the Design of Neolithic Woodworking Tools." Exarc.net.

Ennos, R. 2020. *The Age of Wood*. New York: Scribner.

Ennos, R. 2023. *The Science of Spin*. New York: Scribner.

Farrell, J. W. 2020. *The Clock and the Camshaft*, Guilford, CT: Prometheus Books.

Fernández-Crespo, T. J. Ordoño, F. Etxeberria et al. 2023. "Large-Scale Violence in Late Neolithic Western Europe Based on Expanded Skeletal Evidence from San Juan ante Portam Latinam." *Scientific Reports* 13:17103.

Fleckinger, A. 2018. "Ötzi the Iceman: The Full Facts at a Glance." Czech Republic: Folio Verlagsges. Mbh.

Fressoz, J.-B. 2024. *More and More and More: An All-Consuming History of Energy*. London: Allen Lane.

Gläser, N., B. P. Kneubuehl, S. Zuber, et al. 2011. "Biomechanical Examination of Blunt Trauma due to Baseball Bat Blows to the Head." *Journal of Forensic Biomechanics* 2:F100601,

Goodall, J. 2010. *Through a Window: My Thirty Years with the Chimpanzees of Gombe*. London: Weidenfeld and Nicholson.

Gordon, J. E. 1968. *The New Science of Strong Materials, or Why You Don't Fall Through the Floor*. London: Penguin.

Gordon, J. E. 1978. *Structures, or Why Things Don't Fall Down*. London: Penguin.

Goulson, D. 2019. *The Garden Jungle or Gardening to Save the Planet*. London: Jonathan Cape.

Gowlett, J. A. J. 2016. "The Discovery of Fire by Humans: A Long and Convoluted Process." *Philosophical Transactions of the Royal Society B* 371. https://doi.org/10.1098/rstb.2015.0164.

Gumert, M. D., and S. Malaivijitnond. 2013. "Long-Tailed Macaques Select Mass of Stone Tools According to Food Type. *Philosophical Transactions of the Royal Society B*. 368. 10.1098 /rstb.2012.0413.

Harari, Y. N. 2015. *Sapiens: A Brief History of Humankind*. New York: Harper.

Harmand, S., J. E. Lewis, C. S. Feibel, et al. 2015. "3.3-Million-Year-Old Stone Tools from Lomekwi 3, West Turkana, Kenya." *Nature* 521:310–15.

Henrich, J. 2016. *The Secret of Our Success*. Princeton, NJ: Princeton University Press.

Hernandez-Aguilar, R. A., J. Moore, and T. R. Pickering. 2007. "Savannah Chimpanzees Use Tools to Harvest the Underground Storage Organs of Plants." *Proceedings of the National Academy of Sciences* 104:19210–13.

Herodotus. 2003. *The Histories* London: Penguin Classics.

Hodges, H. 1970. *Technology in the Ancient World.* London: Allen Lane.

Horns, J., R. Jung, and D. R. Carrier. 2015. "In Vitro Strain in Human Metacarpal Bones During Striking: Testing the Pugilism Hypothesis of Hominin Hand Evolution." *Journal of Experimental Biology* 218:3215–21.

Hosseini, K. 2003. *The Kite Runner.* New York: Riverhead Books.

IPBES. 2019. *Global Assessment Report on Biodiversity and Ecosystem Services of the Intergovernmental Science-Policy Platform on Biodiversity and Ecosystem Services.* E. S. Brondizio, J. Settale, S. Diaz, and H. Y. Ngo (eds.), IPBES secretariat, Bonn, Germany.

Jarvis, A. 1985. *Hydraulic Machines.* Aylesbury, UK: Shire Publications.

Jenkins, N. 1980. *The Boat Beneath the Pyramid: King Cheops' Royal Ship.* New York: Holt, Rinehart, and Winston.

Jørgensen, S. 1985. *Tree-Felling with Original Neolithic Flint Axes in Draved Wood. Report on the Experiments in 1952–1954.* Copenhagen: National Museum of Denmark.

Keeley, L., and N. Toth. 1981. "Microwear Polishes on Early Stone Tools from Koobi Fora, Kenya." *Nature* 293:464–65.

Key. A., I. Farr, R. Hunter, A. Mika, M. I. Eren, and S. L. Winter. 2021. "Why Invent the Handle? Electromyography (EMG) and Efficiency of Use Data Investigating the Prehistoric Origin and Selection of Hafted Stone Knives." *Archeological and Anthropological Sciences* 13:162.

Koops, K., A. G. Soumah, K. L. van Leeuwen, H. D. Camara, and T. Matsuzawa. 2022. "Field Experiments Find No Evidence That Chimpanzee Nut Cracking Can Be Independently Innovated." *Nature Human Behaviour* 6:487–94.

Kraft, T. S., V. V. Venkataraman, I. J. Wallace, et al. 2021. "The Energetics of Uniquely Human Subsistence Strategies." *Science* 374. eabf0130.

Landels, J. G. 1978. *Engineering in the Ancient World*. London: Chatto and Windus.

Leakey, M. D. 1971. *Olduvai Gorge, III: Excavations in Beds I and II, 1960–1963*. Cambridge: Cambridge University Press.

Levinson, M. 2016. *The Box: How the Shipping Container Made the World Smaller and the World Economy Bigger*. Princeton, NJ: Princeton University Press.

Logan, W. B. 2005. *Oak: The Frame of Civilization*. New York: W. W. Norton.

Lombard, M., and M. H. Haidle. "Thinking a Bow-and-Arrow Set: Cognitive Implications of Middle Stone Age Bow and Stone-Tipped Arrow Technology." *Archeological Journal* 22:237–64.

Lucas, P. W. 2004. *Dental Functional Morphology: How Teeth Work*. Cambridge: Cambridge University Press.

Lucas, P.W., T. K. Lowrey, B. P. Fereira, V. Sarafis, and W. Kuhn. 1991. "The Ecology of *Mezettia leptoda* (Hk. F. et Thomas) Oliv. (Annonaceae) Seeds as Viewed from a Mechanical Perspective." *Functional Ecology* 5:545–53.

Madsen, B. 1984. "Flint Axe Manufacture in the Neolithic: Experiments with Grinding and Polishing of Thin-Butted Flint Axes." *Journal of Danish Archeology* 3:47–62.

Manning, P. L., D. J. Payne, J. Pennicott, P. M. Barrett, and A. R. Ennos. 2006. "Dinosaur Killer Claws or Climbing Crampons?" *Biology Letters* 2:110–12.

Mathieu, J. R., and D. A. Meyer. 1997. "Comparing Axe Heads of Stone, Bronze, and Steel: Studies in Experimental Archeology." *Journal of Field Archeology* 24:333–51.

McGrail, S. 1996. "The Bronze Age in Europe." In *The Earliest Ships*, edited by R. Gardiner. London: Conway Maritime Press.

McNeil, I. 1972. *Hydraulic Power*. London: Longmans.

McPherron, S. P., Z. Alemsegad, C. W. Marean, et al. 2010. "Evidence for Stone-Assisted Consumption of Animal Tissues Before 3.9 Million Years Ago at Dikika, Ethiopia." *Nature* 466:857–60.

Metz, L., J. E. Lewis, and L. Slimak. 2023. "Bow-and-Arrow, Technology of the First Modern Humans in Europe 54,000 Years Ago at Mandrin, France." *Scientific Advances* 9, eadd4675.

Milks, A., D. Parker, and M. Pope. 2019. "External Ballistics of Pleistocene Hand-Thrown Spears: Experimental Performance Data and Implications for Human Evolution." *Scientific Reports* 25:820.

Mirazón Lahr, M., F. Rivera, R. K. Power, et al. 2016. "Inter-Group Violence Among Early Holocene Hunter-Gatherers of West Turkana, Kenya." *Nature* 529:394–98.

Nicholls, E., A. Ely, L. Birkin, P. Basu, and D. Goulson. 2020. "The Contribution of Small-Scale Food Production in Urban Areas to the Sustainable Development Goals: A Review and Case Study." *Sustainability Science* 15:1585–99.

O'Neill, B., C. Kropp, F. Trommer, and V. Töngi. 2022. "(Re)constructing an Early Medieval Irish Ard." EXARC. https://exarc.net/ark:/88735/10639.

O'Neill, M. C., B. R. Umberger, N. B. Holowka, S. G. Larson, and P. J. Reiser. 2017. "Chimpanzee Super Strength and Human Skeletal Muscle Evolution." *Proceedings of the National Academy of Sciences* 114:7343–48.

Pascual-Garrido, A., S. Carvalho, D. Mjungu, E. Schulz-Kornas, and A. van Casteren. 2025. "Engineering Skills in the Manufacture of Tools by Wild Chimpanzees." *iScience* 112158.

Permana, A. L., J. J. Permana, L. Nellissen, et al. 2024. "The Ontogeny of Nest-Building Behaviour in Sumatran Orang-Utans, *Pongo abelii*." *Animal Behaviour* 214:53–67.

Plummer, T. W., J. S. Oliver, E. M. Finestone, et al. 2023. "Expanded Geographic Distribution and Dietary Strategies of the Earliest Oldowan Hominins and *Paranthropus*." *Science* 379:561–66.

Ponting, C. 2005. *Gunpowder: An Explosive History*. London: Chatto and Windus.

Pontzer, H. 2021. *Burn: New Research Blows the Lid Off How We Really Burn Calories, Lose Weight, and Stay Healthy*. New York: Avery.

Pruetz, J. D., and P. Bertolani. 2007. "Savannah Chimpanzees, *Pan troglodytes verus*, Hunt with Tools." *Current Biology* 17:412–17.

Rasenberger, J. 2020. *Revolver: Sam Colt and the Six-Shooter That Changed America*. New York: Scribner.

Raverat, G. 2018. *Period Piece: A Cambridge Childhood*. London: Faber and Faber.

Reader, J. 2009. *The Untold History of the Potato*. London: Vintage.

Reyssat, E., T. Tallinen, M. Le Merrer, and L. Mahadevan. 2012. "Slicing Softly with Shear." *Physical Review Letters* 109:244301.

Roach, N. T., and D. E. Lieberman. 2014. "Upper Body Contributions to Power Generation During Rapid Overhand Throwing in Humans." *Journal of Experimental Biology* 217:2139–49.

Roach, N. T., M. Venkadesan, M. J. Rainbow, and D. E Lieberman. 2013. "Elastic Energy Storage in the Shoulder and High-Speed Throwing in *Homo*." *Nature* 498:483–87.

Rostain, S., A. Dorison, G. de Saulieu, et al. 2024. "Two Thousand Years of Garden Urbanism in the Upper Amazon." *Science* 383:183–89.

Samson, D. R., A. N. Crittenden, I. A. Mabulla, and A. Z. P. Mabulla. 2017. "The Evolution of Human Sleep: Technological and Cultural Innovation Associated with Sleep-Wake Regulation Among Hadza Hunter-Gatherers." *Journal of Human Evolution* 113:91–102.

Scott, J. 2019. *How the Old World Ended*. New Haven, CT: Yale University Press.

Scott, J. C. 2017. *Against the Grain*. New Haven, CT: Yale University Press.

Skempton, A. W. 1981. *John Smeaton, FRS*. London: Thomas Telford.

Skinner, M. M., N. B. Stephens, Z. J. Tsegai, et al. 2015. "Human-Like Hand Use in *Australopithecus africanus*." *Science* 347:395–97.

Smeaton, J. 1759. "An Experimental Enquiry Concerning the Natural Powers of Water and Wind to Turn Mills and Other Machines, Depending on a Circular Motion." *Philosophical Transactions of the Royal Society* 51:100–174.

Smil, V. 2017. *Energy and Civilization: A History*. Cambridge, MA: MIT Press.

Smith, A. 1776. *The Wealth of Nations.*

Smith, N. 1975. *Man and Water: A History of Hydro-Technology.* London: Charles Scribner's Sons.

Smith, N. 1980. "The Origins of the Water Turbine." *Scientific American* 242:138–48.

Stanley, E., E. Thomson, G. Smith, and K. L. Lamb. 2018. "An Analysis of the Three-Dimensional Kinetics and Kinematics of Maximal Effort Punches Among Amateur Boxers." *International Journal of Performance Analysis in Sport* 18:835–54.

Stapert, D., and L. Johansen. 1999. "Making Fire in the Stone Age: Flint and Pyrite." *Geologie en Mijnbouw* 78:147–64.

Stevenson, R. D., and R. J. Wassersug. 1993. "Horsepower from a Horse." *Nature* 364:195.

Streily, A. H. 2001. "Early Pottery Kilns in the Middle East." *Paléorient* 26:69–81.

Suzman, J. 2022. *Work: A Deep History, from the Stone Age to the Age of Robots.* London: Penguin.

Thieme, H. 1997. "Lower Palaeolithic Hunting Spears from Germany." *Nature* 385:807–10.

Thompson, J. C., S. Carvalho, C. W. Marean, and Z. Alemseged. 2019. "Origins of the Human Predatory Pattern: The Transition to Large-Animal Exploitation by Early Hominins." *Current Anthropology* 60.

Thorpe, K. S., R. H. Crompton, M. M. Gunther, R. F. Ker, and R. McN. Alexander. 1999. "Dimensions and Moment Arms of the Hind- and Forelimb Muscles of Common Chimpanzees (*Pan troglodytes*)." *American Journal of Physical Anthropology* 110:179–99.

Thucidides. 1974. *History of the Peloponnesian War.* London: Penguin Classics.

Tomasso, S., and V. Rots. 2018. "What Is the Use of Shaping a Tang? Tool Use and Hafting of Tanged Tools in the Aterian of Northern Africa." *Archeological and Anthropological Sciences* 10:1389–1417.

Tydgadt, L., and V. Rots. 2022. "Stick to It! Mechanical Performance Tests to Explore the Resilience of Prehistoric Glues in Hafting." *Archaeometry* 64:1252–69.

Van Casteren, A., W. Sellers, S. Thorpe, et al. 2012. "Nest Building Orangutans Demonstrate Engineering Know-How to Produce Safe, Comfortable Beds." *Proceedings of the National Academy of Sciences* 109:6873–77.

Van Schaik, C. P. 2004. *Among Orangutans: Red Apes and the Rise of Human Culture.* Cambridge, MA: Harvard University Press.

Visalberghi, E., and D. Fragaszy. 2013. "The EthoCebus Project. Stone Tool Use by Wild Capuchin Monkeys." In *Multidisciplinary Perspectives on the Cognition and Ecology of Tool Using Behaviors,* edited by C. Sanz, J. Call, and C. Boesch. Cambridge: Cambridge University Press.

Warde, P. 2007. *Energy Consumption in England and Wales, 1560–2000.* Rome: Consiglio Nazionale Delle Ricerche.

Westcott, D. 1999. *Primitive Technology: A Book of Earth Skills.* Salt Lake City, UT: Gibbs-Smith.

Wilson, C. 1965. *England's Apprenticeship,* London: Longmans.

Wilson, M. L., C. Boesch, B. Fruth, et al. 2014. "Lethal Aggression in *Pan* Is Better Explained by Adaptive Strategies Than Human Impacts." *Nature* 513:7518.

Winchester, S. 2018. *Exactly: How Precision Engineers Created the Modern World.* London: William Collins.

Wrangham, R. 2009. *Catching Fire: How Cooking Made Us Human.* London: Profile Books.

Wrigley, E. A. 2010. *Energy and the English Industrial Revolution.* Cambridge: Cambridge University Press.

Yoganandan, N., F. A. Pintar, A. Sances, P. R. Walsh, C. L. Ewing, and D. J. Thomas. 1995. "Biomechanics of Skull Fracture." *Journal of Neurotrauma* 12:659–68.

Young, R. W. 2003. "Evolution of the Human Hand: The Role of Throwing and Clubbing." *Journal of Anatomy* 202:165–74.

Zanette, L. Y., N. R. Frizzelle, M. Clinchy, et al. 2023. "Fear of the Human 'Super Predator' Pervades the South African Savannah." *Current Biology* 33:4689–96.

Index